T0137428

Applications of Soft Computing for the Web

Applications of Soft Computing for the Web

Rashid Ali · M. M. Sufyan Beg
Editors

Applications of Soft Computing for the Web

 Springer

Editors
Rashid Ali
Department of Computer Engineering
Aligarh Muslim University
Aligarh, Uttar Pradesh
India

M. M. Sufyan Beg
Department of Computer Engineering
Aligarh Muslim University
Aligarh, Uttar Pradesh
India

ISBN 978-981-13-4994-2   ISBN 978-981-10-7098-3   (eBook)
https://doi.org/10.1007/978-981-10-7098-3

© Springer Nature Singapore Pte Ltd. 2017
Softcover re-print of the Hardcover 1st edition 2017
This work is subject to copyright. All rights are reserved by the Publisher, whether the whole or part of the material is concerned, specifically the rights of translation, reprinting, reuse of illustrations, recitation, broadcasting, reproduction on microfilms or in any other physical way, and transmission or information storage and retrieval, electronic adaptation, computer software, or by similar or dissimilar methodology now known or hereafter developed.
The use of general descriptive names, registered names, trademarks, service marks, etc. in this publication does not imply, even in the absence of a specific statement, that such names are exempt from the relevant protective laws and regulations and therefore free for general use.
The publisher, the authors and the editors are safe to assume that the advice and information in this book are believed to be true and accurate at the date of publication. Neither the publisher nor the authors or the editors give a warranty, express or implied, with respect to the material contained herein or for any errors or omissions that may have been made. The publisher remains neutral with regard to jurisdictional claims in published maps and institutional affiliations.

Printed on acid-free paper

This Springer imprint is published by Springer Nature
The registered company is Springer Nature Singapore Pte Ltd.
The registered company address is: 152 Beach Road, #21-01/04 Gateway East, Singapore 189721, Singapore

*To Prof. L. A. Zadeh, who left so much for us
before he left us*

Foreword

A debate centered on *for and against* the utility of soft computing in solving some of the real-life problems has reached a steady state. Soft computing, a term coined by Prof. Lotfi Zadeh, the father of fuzzy logic, is also sometimes referred as computational intelligence and could be broadly classified into two categories: 1. human-centric and 2. animal-centric. The first category includes computing with words via fuzzy sets and fuzzy logic, rough sets, soft sets, neural networks, genetic algorithms, support vector machine, while animal-/insect-centric category comprises of swarm optimization, ant colony optimization, anticipation, and alike. Steady increase in publication of research articles is a testimony for wide acceptance of soft computing which is Out *of the Box Solution in data analytics.*

Professor M. M. Sufyan Beg, a brilliant scholar of eminence, and his colleague, Dr. Rashid Ali, rightly thought of publishing a unique book comprising a total of 11 sections dealing with Internet and Web-based applications of soft computing techniques.

The first section contains a summary of soft computing based recommender systems with focus on genetic algorithms, similarity measures, and linguistic quantifiers. It also provides the enhanced multicriteria systems based on analytic hierarchy process. Online summarization of documents, which can be proved to be very effective tool for the software industry especially, and online clustering of documents are discussed in subsequent sections. Data warehousing, data mining, and more generally Web/cloud based in many different formats is the order of the day. There is a need for effective, efficient, and speedy algorithms to search, mine, and extract the data in the required format. The section on Web extraction tools carefully deals with this topic. Uncertainty in weather prediction calls for sustained research efforts to maximize the profit ratio in minimal human efforts in agriculture-based products. The research articles in this book carefully deal with this important topic.

Human health is the most important part in all the era. Development of expert systems using various techniques is in process and needs all-time attention due to change in conditions like lifestyle, behavioral changes in the microorganisms, effects of global warming, changes in nature on the health of human being. The book has an important section dealing with Web-based medical diagnosis based on several years of experience of the researchers.

Research effort dealing security issues is the need of the changing world. For variety of reasons, Web security assumes significant importance. Popularity of e-marketing and e-commerce cannot be underestimated. The papers in this document on business intelligence are a step forward in right direction. Part VIII gives the intelligent system for online marketing.

In the world of Internet, Web, cloud, the major problem that can occur is network traffic. Best possible way to use the available bandwidth and optimize network usage is to implement effective ways to achieve Quality of Service (QoS). It can be used in wired or wireless network to enhance the quality of network service and to get the maximum network speed with minimum collision. The book also presents articles on QoS and enhancement of network traffic.

I am amazed to see that the authors tried to cover variety of soft computing Web-based applications which will interest the readers, but research is endless. Therefore, there is a separate section on emerging applications of soft computing techniques. I congratulate all the authors for their excellent articles.

Professor Sufyan Beg and his team have taken endless efforts in bringing out the excellent edited book for which they deserve a loud applause. I wish Prof. Beg and his team a grand success in their further academic endeavors.

Professor Ashok Deshpande, Ph.D. (Engineering)
Founding Chair: Berkeley Initiative in Soft Computing
(BISC)–Special Interest Group (SIG)–Environment
Management Systems (EMS)
Guest Faculty: University of California, Berkeley, CA, USA
Visiting Professor: University of New South Wales,
Canberra, Australia, Essen University, Germany, and IITB, India
Adjunct Professor: NIT, Silchar, and COEP, Pune
Former Deputy Director: NEERI/CSIR, India

Preface

Broadly speaking, soft computing encompasses unorthodox computing techniques like fuzzy logic, artificial neural networks, evolutionary computation, and nature-inspired computing. The intrinsically imprecise nature of the human beings can best be mimicked with soft computing techniques. The perceptions of height being tall or short, colors being bright or dark, appearance being pretty or ugly, weather being hot or cold, and so on, require shades and can't be attributed with a precise value. This highly subjective nature of perception is best dealt with the notion of fuzzy logic. The way human brain works with the estimates and not the precise values, necessitates the adoption of artificial neural networks for solving the real-life problems. The way living beings evolve into better beings, under a given objective of being "better," is adopted under the premises of evolutionary computation. Other nature-inspired computing techniques like swarm Intelligence, ant colony optimization, simulated annealing try to solve the real-life problems the way the nature solves its complex problems.

In an interesting post dated November 25, 2010, Prof. L. A. Zadeh, the Father of Fuzzy Logic, has defined "soft computing" as follows:

> Soft computing is a coalition of methodologies which collectively constitute a system with wide-ranging capabilities for the conception, design, and operation of intelligent systems. The principal members of the coalition are: fuzzy logic, neurocomputing, evolutionary computing, and probabilistic computing. The guiding principle of soft computing is: In general, better results can be achieved through the use of constituent methodologies of soft computing in combination rather than isolation. In recent years, the label "Computational Intelligence" has gained popularity. The meaning of computational intelligence is close to the meaning of soft computing.

Professor Zadeh has also defined Fuzzy Logic in another post dated December 11, 2008, as follows:

> Fuzzy logic is a precise system of reasoning, deduction, and computation in which the objects of discourse and analysis are associated with information which is, or is allowed to be, imperfect. Imperfect information is defined as information which in one or more respects is imprecise, uncertain, vague, incomplete, partially true, or partially possible. In

fuzzy logic everything is or is allowed to be a matter of degree. Degrees are allowed to be fuzzy. Fuzzy logic is not a replacement for bivalent logic or bivalent-logic-based probability theory. Fuzzy logic adds to bivalent logic and bivalent-logic-based probability theory, a wide range of concepts and techniques for dealing with imperfect information. Fuzzy logic is designed to address problems in reasoning, deduction, and computation with imperfect information which are beyond the reach of traditional methods based on bivalent logic and bivalent-logic-based probability theory. In fuzzy logic, the writing instrument is a spray pen with precisely known adjustable spray pattern. In bivalent logic, the writing instrument is a ballpoint pen. The importance of fuzzy logic derives from the fact that in much of the real world imperfect information is the norm rather than exception.

Now, when it comes to mine something as human-generated and uncontrolled as the World Wide Web, we must be looking for tools to counter as savage the imprecision, ambiguity, and uncertainty as possible. The Web is an ever-growing phenomenon. A problem thus arises in exploring the Web for bits and pieces of information. The search may vary from a text-based query to multimedia searches and also to clue-based searching. Leave alone the multimedia searches or the clue-based searching, even the simple keyword search is quite different from the *Find* facility in file managers/editors. This difference arises not only due to the huge volume of data distributed at so many physically spaced sites, but also due to the typical pattern of organizing the data in these sites. A hierarchical pattern of organizing the data is generally followed, with each document having a home page as its main module, with many hyperlinks to the related pages or even to other Web sites. This pattern continues till the last level of pages called terminal pages or leaf pages. Such a pattern not only keeps the retrieval latency low by virtue of providing the information in smaller units, but also renders levels of abstraction to the information, which is very essential in the structured approach. It is for such an organizational pattern of information that the resource organization and discovery is to be employed on the Web. This is where soft computing fits the requirements very well.

The advent of social media has given the Web an altogether new dimension. Since the social media is all about the human-to-human interactions, the soft computing techniques seem to be the only plausible solution to cater to the needs of mining it. Similarly, the recommender system has to be as human-like as possible. The soft computing techniques can play their role effectively there too. The very notion of optimization in areas like quality of service makes a strong case for the application of Genetic Algorithms. In the literature, fuzzy logic has been extensively employed to reach consensus in the generic domain of multicriteria group decision making. So, online decisions for issues like crop selection or gynecological disease diagnosis necessitate the use of fuzzy logic.

With the above diverse issues in mind, we have thought of bringing this edited book out. We have no hesitation in admitting that there is so much more still left to be explored in the area of soft computing vis-à-vis the World Wide Web. With this humble beginning, however, we hope to trigger a series of discussions on this topic.

The editors are thankful to all the authors who have contributed excellently in the preparation of this book. We are grateful to Prof. Ashok Deshpande for having agreed to write a foreword for his book. We are also thankful to all the reviewers, who have done their job diligently and within the deadline posed to them. The help from the publishers' team is also worth a special mention. We would like to specially thank Anjana Bhargavan (Ms.), Project Coordinator, Books Production, Springer Nature, for cooperation and support, Yeshmeena Bisht, Consultant and Suvira Srivastav, Associate Editorial Director, Computer Science & Publishing Development, Springer India, for giving us an opportunity to publish our book with Springer Nature. Last but not the least, we are thankful to our family members who have been patient and supportive of this project.

Aligarh, India
Rashid Ali
M. M. Sufyan Beg

Contents

About the Editors

Dr. Rashid Ali obtained his B.Tech. and M.Tech. from Aligarh Muslim University, India, in 1999 and 2001, respectively. He went on to acquire his Ph.D. in Computer Engineering from the same university in 2010 and is currently working as an Associate Professor in the Department of Computer Engineering, Aligarh Muslim University, India. He has authored 8 book chapters and over 100 papers in various international journals and international conference proceedings of repute. His research interests include Web Searching, Web Mining, and Soft Computing Techniques.

Prof. M. M. Sufyan Beg holds a B.Tech. from Aligarh Muslim University, M.Tech. from the Indian Institute of Technology Kanpur (IIT Kanpur), India, and a Ph.D. from the Indian Institute of Technology Delhi (IIT Delhi), India. Currently, he is a Professor in the Department of Computer Engineering, Aligarh Muslim University, India. He has also visited the University of California, Berkeley, as a BT Fellow. He has published 16 book chapters and over 115 papers. He has served as a reviewer and on the editorial board of many prominent journals. His current research interests are in the areas of Soft Computing, Question Answering Systems, Web Mining and Searching.

List of Reviewers

1. Dr. M. Tanveer, Ramanujan Fellow, Discipline of Mathematics, Indian Institute of Technology Indore, India
2. Deepak K. Bagchi, Head, Smart Data, SLK Software Services Pvt. Ltd., Bengaluru, India
3. Vibhor Kant, Assistant Professor, Department of Computer Science and Engineering, LNM Institute of Information Technology, Jaipur, India
4. Dr. Tasleem Arif, Head, Department of Information Technology, Baba Ghulam Shah Badshah University, Rajouri, Jammu and Kashmir, India
5. Dr. Mohammad Nadeem, Assistant Professor, Department of Computer Science, Aligarh Muslim University, Aligarh, India
6. Abdul Rahman, Assistant Professor, Department of Computer Science, Faculty of Technology, Debre Tabor University, Debre Tabor, Ethiopia.
7. Dr. Bindu Garg, Dean, R&D, Bharati Vidyapeeth College of Engineering, New Delhi, India
8. Dr. B. M. Imran, Professor, Faculty of CCSIT, University of Dammam, Kingdom of Saudi Arabia
9. Dr. Majid Bashir Malik, Sr. Assistant Professor, Department of Computer Science, Baba Ghulam Shah Badshah University, Rajouri, Jammu and Kashmir, India
10. Dr. Jatinder Manhas, Sr. Assistant Professor, I/C Head, Computer Sciences & IT (BC), University of Jammu, Jammu, Jammu and Kashmir, India
11. Dr. Abid Sarwar, Assistant Professor, Department of Computer Science & IT, University of Jammu (BC), Jammu, Jammu and Kashmir, India
12. Dr. Gulfam Ahamad, Department of Computer Science & Information Technology, School of Computer Science & Information Technology, Maulana Azad National Urdu University (Central University), Hyderabad, India
13. Dr. Mohd Ashraf, Associate Professor, Computer Science Engineering Department, Polytechnic, Maulana Azad National Urdu University (Central University), Hyderabad, India

14. Dr. Mohd Naseem, Assistant Professor, Department of Computer Science and Engineering, Thapar University, Patiala
15. Dr. Q. M. Danish Lohani, Assistant Professor, Department of Mathematics, South Asian University, New Delhi, India
16. Sayyad Mohd Zakariya, Assistant Professor, Electrical Engineering Section, University Polytechnic, Aligarh Muslim University, Aligarh, India
17. Dr. Mohammad Shahid, Assistant Professor (Computer Science), Department of Commerce, Aligarh Muslim University, Aligarh, India
18. Musheer Ahmad, Assistant Professor, Department of Computer Engineering, F/O Engineering and Technology, Jamia Millia Islamia (Central University), New Delhi, India
19. Dr. Yacine Laalaoui, Assistant Professor, IT Department, College of Computers and Information Technology, Taif University, Taif, Kingdom of Saudi Arabia

Introduction

Rashid Ali and M. M. Sufyan Beg

1 Introduction

The World Wide Web (in short, the Web) is a huge collection of dynamic, diverse, and heterogeneous documents. Since its inception, the Web has grown at an explosive rate. Users have become more involved in the web-based activities, and the web resources are being utilized a lot. The Internet has become a place for variety of actions and events related to work, business, health care, social interactions, education, and entertainment. For example, online social network sites are being heavily used by millions of users daily for information sharing, interaction, propagation, etc., which further increases the amount of data available on the Web. Now, size of the Web is so large that even best of the search engines cannot index the complete web effectively. Moreover, the data stored on the Web is mostly unlabeled, semi-structured, distributed, dynamic, and heterogeneous. This makes the finding of information from the Web, which is specific to a user's need, tough because of the subjectivity, vagueness, uncertainty, context sensitivity, and imprecision present therein. As a result, the user's need for the desired information is not fulfilled satisfactorily. The situation is such that we are sailing through an ocean (of information) but not finding a single cup of water (knowledge) to quench our thirst.

Therefore, developments of web-based applications which can incorporate human's intelligence, filter out irrelevant information, search through nonstructural data, and provide suitable answers to user's questions as per their need in different fields of e-commerce, e-business, e-health, adaptive web systems, and information

R. Ali (✉) · M. M. S. Beg
Department of Computer Engineering, Z. H. College of Engineering
and Technology, Aligarh Muslim University, Aligarh 202002, Uttar Pradesh, India
e-mail: rashidaliamu@rediffmail.com

M. M. S. Beg
e-mail: mmsbeg@eecs.berkeley.edu

© Springer Nature Singapore Pte Ltd. 2017
R. Ali and M. M. S. Beg (eds.), *Applications of Soft Computing for the Web*,
https://doi.org/10.1007/978-981-10-7098-3_1

retrieval are needed. For example, in e-commerce, recommender systems should suggest suitable products to customers or provide some useful information to help them decide which items to buy. In e-business, system should learn customer needs correctly, identify future trends, and in the long run increase customer's loyalty. In e-health, the system should learn a patient's health conditions accurately and provide proper and effective e-health solutions. In adaptive web systems, the organization and presentation of the website should be customized effectively to match the precise needs of visitors. In information retrieval, user's information need should be learnt accurately and most appropriate results should be presented before the user.

But, the challenge is to devise effective methods to learn the user's interest, need, or preference from the uncertain or ambiguous web data and to design intelligent and effective web-based systems. In this context, soft computing is a natural choice as the soft computing techniques are well known for their tolerance for imprecision, uncertainty, vagueness, approximations, and partial truth.

In the past, soft computing techniques have been successfully applied for many applications on the Web. In this book, we present chapters on recent efforts made in the area of soft computing applications to the Web. This chapter provides a brief introduction of the different chapters of the book.

In the next section, we present a brief overview of major soft computing techniques. Thereafter, we describe briefly the work presented in different chapters of the book.

2 Soft Computing Techniques—An Overview

The term soft computing was quoted by Zadeh in 1990s [1]. Soft computing is a collection of methodologies which work synergistically to provide acceptable solutions to real-life ambiguous problems. The goal is to exploit the tolerance for imprecision, uncertainty, vagueness, approximations, and partial truth in order to achieve tractability, robustness, and low-cost solutions [2]. These techniques have close resemblance to human-like decision-making. The main advantage of the use of soft computing is that they can solve several real-life problems for which there is no mathematical model available.

Soft computing includes topics of computational intelligence, natural computation, and organic computing [3]. The major components of soft computing are fuzzy logic, artificial neural network, and probabilistic reasoning, including belief networks, evolutionary computing, chaos theory, and parts of learning theory [4]. Nowadays, bio-inspired computing including swarm intelligence is also considered as soft computing techniques. Here, we discuss briefly some of the important components of soft computing.

2.1 Fuzzy Logic

Zadeh in 1965 [5] introduced the concept of fuzzy sets which generalizes the concept of crisp set and is well suited for handling the incomplete or imprecise information. Fuzzy sets allow items to have partial degree of membership (i.e., result in the interval [0, 1]) instead of hard membership of the items to the crisp set (i.e., 0 if item does not belong to the set, and 1 for otherwise). Fuzzy logic mimics the human decision-making behavior and is based on fuzzy sets. A detailed discussion on fuzzy sets and fuzzy logic can be found in [6].

2.2 Artificial Neural Networks

Artificial Neural Networks (ANN, in short) is developed to mimic the human brain intelligence and discriminating power. A simple ANN model contains two layers, input layer and output layer of artificial neurons. Artificial neurons mimic the biological neurons which are very simple processing elements in the brain, connected with each other in a massively parallel fashion [7]. Back Propagation Network (BPN) contains an additional intermediate hidden layer of artificial neurons in between input and output layers. ANN is trained by adjusting the weights of the connections between artificial neurons in ANN with the help of training data. The training of ANN is done in such a way that a specific input information leads to a specific target output. Back Propagation (BP) algorithm is one of the popular training algorithms designed to train the neural networks and it minimizes the deviation between the actual and desired objective function values. A detailed discussion on artificial neural networks can be found in [8].

2.3 Genetic Algorithm

Genetic Algorithm (GA, in short) is a stochastic search technique which can provide solutions for diverse optimization problems. GA works on Darwin's principle of "survival of the fittest" and gives a method to model biological evolution. It starts with a population of individuals called chromosomes, where each chromosome has a number of genes and used to decide the possibility of reproduction for the next generation on the basis of its fitness value. The chromosome which has best (fittest) value will be used to form new population iteratively. The initial set of chromosomes is selected for the new generation using selection operator based on their fitness values. Then, crossover is used to generate two new off-springs by swapping a portion of the two parent chromosomes. Mutation operator is used randomly to alter the status of a gene bit and it introduces diversity in the population. Genetic algorithm terminates when a predefined number of generations reached or fitness threshold is met. A detailed discussion on genetic algorithm can be found in [9].

2.4   Swarm Intelligence

Swarm Intelligence (SI, in short) [10] is used to solve complex optimization problems through the principles of self-organization, decentralization, and cooperation through communication. SI concept is inspired from the swarm of social organisms like birds, fish, and animals. There are many different types of SI techniques out of which Particle Swarm Optimization (PSO) and Ant Colony Optimization (ACO) are two important techniques. A detailed discussion on swarm intelligence can be found in [11].

3   Organization of the Book

This book encompasses fourteen chapters devoted to the recent developments in the field of the soft computing tools applied on the Web. We give a special attention to the use of the soft computing tools to solve the problems associated with the online recommender systems. First four chapters of this book discuss soft computing based recommender systems.

In chapter entitled "Context Similarity Measurement Based on Genetic Algorithm for Improved Recommendations", the authors investigate the use of genetic algorithm in conjunction with fuzzy logic for solving the problems associated with collaborative filtering based recommender systems. Authors propose an improved method for identifying the similar preferences of users in two phases. In the first phase, they extend traditional recommender systems by incorporating contextual information into fuzzy collaborative filtering using a contextual rating count approach. In the second phase, a genetic algorithm is employed to learn actual user profile features weight and then, similarity among users in a sparse dataset is computed.

In chapter entitled "Enhanced Multi-criteria Recommender System Based on AHP", the authors propose a multi-criteria recommender system based on AHP (Analytic Hierarchy Process) for college selection. The proposed method builds an analytic hierarchy structure of multiple criteria and alternatives to ease the decision-making process. The system computes an overall score for each college based on multiple criteria and recommends the colleges on the basis of their score.

In chapter entitled "Book Recommender System Using Fuzzy Linguistic Quantifiers", the authors propose a book recommender system that uses Ordered Weighted Aggregation (OWA), a well-known fuzzy aggregation operator to aggregate the ranking of the books obtained from different top-ranked universities and recommends the books on the basis of this overall ranking.

In chapter entitled "Use of Soft Computing Techniques for Recommender Systems: An Overview", authors discuss briefly various efforts made in the area of soft computing based recommender systems. The chapter also points out several possible research directions in the area.

The next section discusses soft computing based online documents summarization. In chapter entitled "Hierarchical Summarization of News Tweets with Twitter-LDA", the author proposes to use Twitter-LDA, a variation of latent drichlet allocation to summarize multi-topic tweets data of verified twitter news source and their social interaction with users. The summarization is done according to time, location, person, or events. The proposed summarization application extracts and mine tweets for each identified topic based on their content and social interaction.

The next section discusses soft computing based web data extraction. In chapter entitled "Bibliographic Data Extraction from the Web Using Fuzzy-Based Techniques", the authors discuss the importance and effectiveness of fuzzy and hybrid string matching techniques for the extraction of publications data from the publication databases and digital libraries. They propose a tool to extract bibliographic data from the Web.

The next section discusses soft computing based question answering systems. In chapter entitled "Crop Selection Using Fuzzy Logic-Based Expert System", the authors discuss fuzzy logic-based expert system to assist farmers in crop selection on the basis of climatic conditions and soil properties. The system suggests the most suitable crop for the given climatic conditions and soil properties.

The next section covers the discussion on soft computing based online healthcare systems. In chapter entitled "Fuzzy Logic Based Web Application for Gynaecology Disease Diagnosis", the authors propose a system for accurate disease diagnosis in the presence of uncertainty, fuzziness, and ambiguity in the complaints to disease entity. The authors try to simulate the process of differential diagnosis using fuzzy logic-based formalisms.

The next section discusses soft computing based online documents clustering. In chapter entitled "An Improved Clustering Method for Text Documents Using Neutrosophic Logic", the authors propose an improved fuzzy c-means clustering algorithm based on neutrosophic logic. The authors propose to use indeterminacy factor of neutrosophic logic to enhance the effectiveness of fuzzy c-means algorithm for text document clustering.

The next section discusses soft computing based web security applications. In chapter entitled "Fuzzy Game Theory for Web Security", the authors propose to use static fuzzy game theory approach to deal with the web application security threats. A non-cooperative game is discussed to demonstrate attacker and defender roles and fuzzy game theoretic approach is proposed based on the interactions between the security guard and the ubiquitous attacker.

The next section discusses soft computing based online market intelligence. In chapter entitled "Fuzzy Models and Business Intelligence in Web-Based Applications", the authors discuss the fuzzy models, their applicability, and ability to provide solutions to different business applications. The authors discuss the use of fuzzy logic for business decision-making with vague data, in the presence of outliers.

The next section consists of two chapters covering the discussion on soft computing based Internet of things applications. In chapter entitled "GSA-CHSR: Gravitational Search Algorithm for Cluster Head Selection and Routing in Wireless Sensor Networks", the authors try to solve the two most promising problems in Wireless Sensor Networks (WSNs), namely cluster head selection and routing using an optimization algorithm called gravitational search algorithm. They develop Cluster Head (CH) selection strategy with an efficient encoding scheme by formulating a novel fitness function based on residual energy, intra-cluster distance, and CH balancing factor. They also devise a Gravitational Search Algorithm (GSA)-based routing algorithm by considering residual energy and distance as optimization parameters.

In chapter entitled "Utilizing Genetic-Based Heuristic Approach to Optimize QOS in Networks", the author proposes to solve routing-related issues using heuristic-based genetic algorithm. The proposed algorithm finds the fittest path from source to destination and optimizes various QOS parameters like hop count, delay, throughput, etc.

The next section discusses an emerging soft computing technique. In chapter entitled "V-MFO: Variable Flight Mosquito Flying Optimization", the author presents a novel optimization algorithm called Variable flight Mosquito Flying Optimization (V-MFO). The proposed algorithm mimics the behavior of mosquitoes to find a hole or an irregularity in a mosquito net.

Finally, we conclude the book in the last chapter.

References

1. Zadeh LA (1994) Fuzzy logic, neural networks, and soft computing. Commun ACM 37 (3):77–84
2. Pal SK, Talwar V, Mitra P (2002) Web mining in soft computing framework: relevance, state of the art and future directions. IEEE Trans Neural Netw 13(5):1163–1177
3. Soft computing home page—short definition of soft computing. Available at www.soft-computing.de. Last accessed on 15 Sept 2017
4. BISC Program; Soft computing people. Available at https://people.eecs.berkeley.edu/~zadeh/. Last accessed on 15 Sept 2017
5. Zadeh LA (1965) Fuzzy sets. Inf Control 8(3):338–353
6. Klir GJ, Yuan B (1995) Fuzzy sets and fuzzy logic: theory and applications. Prentice Hall Inc., Upper Saddle River. ISBN: 0131011715
7. Tettamanzi A, Tomassini M (2013) Soft computing: integrating evolutionary, neural, and fuzzy systems. Springer Science & Business Media, Heidelberg
8. Haykin S (1998) Neural networks: a comprehensive foundation, 2nd edn. Prentice Hall Inc., Upper Saddle River. ISBN: 0132733501
9. Sivanandam SN, Deepa SN (2007) Introduction to genetic algorithms, 1st edn. Springer Publishing Company, New York. ISBN: 354073189X 9783540731894

10. Grosan C, Abraham A, Chis M (2006) Swarm intelligence in data mining. In: Grosan C, Abraham A, Chis M (eds) Swarm intelligence in data mining, studies in computational intelligence. Springer, Berlin, pp 1–20
11. Kennedy J, Eberhart RC (2001) Swarm intelligence. Morgan Kaufmann Publishers Inc., Burlington. ISBN: 1558605959

Part I
Soft Computing Based Recommender Systems

Context Similarity Measurement Based on Genetic Algorithm for Improved Recommendations

Mohammed Wasid and Rashid Ali

1 Introduction

With the rapid growth of the information available on the internet and access to these diverse information creates problems to the Web users to select desirable information. In order to get desirable information, users have to spend more time and energy. Time and energy constraint is a big challenge in front of the online customers. So there is a need for personalization tools that can help Web users to get relevant information easily within constant time. During the past several years, many recommender systems have been developed and keep evolving rapidly. Usually, RSs can be classified into four different categories, namely content-based (CB), collaborative filtering (CF), demographic filtering (DMF), and hybrid systems [1].

Currently, collaborative filtering [2, 3] is the most popular strategy used for personalization and to deal with information overload problem. The general aim of CF is to find similar users to the target user on the basis of similar tastes and preferences liked in the past. In the past few years, CF widely deployed in industry and studied in academia, such as Amazon and Netflix because of its simplicity and efficacy. There are three main steps involved in CF recommendations which are as follows:

- **Data collection**: The first step is the collection of the data for user profile creation. Data can be collected through implicit, explicit, or both methods.

M. Wasid (✉) · R. Ali
Department of Computer Engineering, Aligarh Muslim University,
Aligarh 202002, Uttar Pradesh, India
e-mail: erwasid@gmail.com

R. Ali
e-mail: rashidaliamu@rediffmail.com

© Springer Nature Singapore Pte Ltd. 2017
R. Ali and M. M. S. Beg (eds.), *Applications of Soft Computing for the Web*,
https://doi.org/10.1007/978-981-10-7098-3_2

- **Similarity computation**: After profile creation in the previous step, the CF computes similarity between the target user and other remaining users.
- **Prediction and recommendation**: Finally, predicted items recommended to the target user.

Although CF has emerged as the most prevalent method, it suffers from the following major flaws [1].

- **Data sparsity**: In traditional CF, a rating matrix contains millions of the users and items. It is quite difficult for a user to provide ratings to all available items as she might not have used all those items, which leads to the sparsity problem and plays the negative role while computing the similarity among the users.
- **Cold start problem**: When a new user or item comes into the system, these items cannot be recommended to anyone until it has collected a substantial number of ratings given by users in the systems. This problem referred to as cold start problem and can be alleviated by hybrid filtering scheme, where the CB system is combined with CF system.

The CF method focuses mainly on user-provided explicit item ratings while computing the similarity between users and items. But it is very difficult to compute similarity in case of sparse dataset because every user provides a very limited number of item ratings compared to the large itemset available in the system. To deal with such issues, other user–item-related information (age, gender, genre, occupation, context, etc.) can be incorporated in order to support the effective similarity computation. But after attaching these user–item features, one more thing that we have to take into consideration is the uncertainty associated with user–item features. Fuzzy logic [4] seems to be an ideal choice for handling the uncertainty issue associated with these features. Therefore, we use fuzzy CF recommender systems that fuzzify the age and genre features. Fuzzy logic gives high-value properties to manage uncertainty in a unified way and also have the aptitude to perform rough reasoning. Fuzzy logic can help to minimize many usual problems from which current recommender systems are suffering.

The concept of context awareness in collaborative filtering (CACF) [5] was introduced to deal with the situation-based recommendations and to retrieve the most relevant information to the user. CACF can be used to generate any location, mood, and time-dependent collaborative recommendations [6]. For example, a user is depressed and wants to watch a motivational movie in the theater to change his mood or user want to purchase a gift for her friend for his birthday on weekend. Therefore, by seeing these examples, we can say that contexts play a key role in making powerful recommendation for the user. Generally, there are three models used to generate context-based recommendations [7].

- **Contextual pre-filtering**: Initially, this approach applies certain context-dependent criteria to filter out data and then select a particular item on the given contextual condition.

- **Contextual post-filtering**: This approach is applied after completion of some traditional two-dimensional recommendation technique.
- **Contextual modeling**: This approach directly incorporates the contextual information into recommendations algorithms.

Sometimes, when we choose a set of contextual attributes which may not provide accurate predictions, therefore, we choose only those optimal set of contextual attributes which generate the best neighborhood set for appropriate recommendations. Genetic algorithm (GA) [8] operation is based on the Darwinian principle of "survival of the fittest" which try to utilize artificial evolution to get improved results in each generation. The process of GA starts with an initial population and consists of chromosome or genotype, where each chromosome has its own fitness value which is used to determine the superiority of the old and newly generated chromosome in the competition. Furthermore, two operators, namely crossover and mutation, are used to create new individuals, where crossover creates two new individuals by using two parent chromosomes through exchanging the needed information and mutation allows maintaining the genetic properties of the individuals by allowing some minor changes to the genes. This procedure iterates for some time till a predefined number of iterations or threshold reached.

The main contribution of our work is employing the genetic algorithm to learn feature weights of the hybrid user profile. Each item's rating is then predicted based on these features. Our proposed method is different from previous attempts at measuring contextual similarity, as the user profile features not only considered the overall user ratings but also the user–item background information. We presented an experimental analysis on LDOS-CoMoDa dataset, and improving the quality of the recommendation using genetic algorithm is one of the motivations behind this work.

The remainder of this paper is outlined as follows. Section 2 introduces the overview of collaborative filtering and context awareness. The genetic algorithm-based contextual recommender system is described in Sect. 3. The experimental results and comparison of proposed method with previous methods are presented in Sect. 4. We conclude the paper with possible future direction of this work in Sect. 5.

2 Literature Review

In the literature, there has been an extensive study on recommender systems and most of the studies used user-rated overall ratings for recommendation purpose. In this section, we provide an overview of several major sets of approaches for the suitability of our proposed framework.

As we discussed in the previous section, recommender systems provide recommendation to the user based on multiple filtering techniques but broadly, it is divided into collaborative filtering [2, 3] and content-based filtering [9] techniques, where CF works on the basis of similar like-minded users, i.e., an item will be recommended to a user if her similar users liked it in the past. Examples of such techniques that find like-minded users may cover nearest-neighbor approach [10], restricted Boltzmann machine [11], matrix completion [12], Bayesian matrix factorization [13], etc. The CF approach further classified into user collaborative filtering and item collaborative filtering [3]. User-based collaborative filtering [10] computes the similarity among the users based on the items they preferred. Then, the rating of target user's unseen item is predicted by using the combined ratings of top similar users, whereas item-based collaborative filtering [14] computes the similarity among items based on the users who preferred both items and then recommend the user those items which she preferred in the past. These two types of CF can also be combined to form user–item-based CF, which generate recommendations based on the user–item matrix by finding a common space for users and items. Matrix factorization techniques [12, 13] can be considered as the examples of user–item-based CF technique. However, CF is the widely used recommendation technique but it also suffers from multiple problems like data sparsity, cold start problem which is often referred to as new user and new items problems.

The second most popular recommendation approach is content-based filtering [9]. This approach provides recommendation by extracting the features from the user or item profile. The base assumption of CB filtering is that similar users will like items similar to the items which they preferred in the past. Linden et al. [15] proposed a method to generate a search query with some item feature that the user preferred in the past to find other similar items for recommendations. Similarly, recommender system builds profiles of users' news likeness based on their past news clicks and provide recommendations to all those users who share the same location [16]. Therefore, in these examples, we can see that CB approach does not rely on the user-provided ratings rather it only checks the user's feature (behavior) and item's features (attributes). So, in case new items come to the system, then based on the item's feature, recommendations can be made to the user who liked the similar feature items in the past. But sometimes it is very difficult to extract user–item features and content-based approach failed. Therefore, researchers are trying to develop new approaches that combine both the CF and CB approaches [16]. Authors in [17–19] developed a set of hybrid features that include some of the users' and items' features and then applied CF approach for recommendations. Melville et al. [20] proposed an effective framework for combining CB and CF and used content-based predictor to enhance existing user data and then provided recommendations through CF approach. Restricted Boltzmann machine was also used to learn similarity between items and then incorporated with collaborative filtering [21].

In most of the cases, the information about the user–item is incomplete and uncertain. How to incorporate fuzzy logic [4] concept into the recommendation systems is an important and interesting problem, which is worth paying attention too. Fuzzy logic is being effectively used in many domain areas by many researchers, several directions are yet to be explored, and one of them is the recommender systems. Tamayo [22] presented and discussed the model and main features of fuzzy recommender system's engine. Thong [23] enhanced the accuracy of medical diagnosis by presenting an intuitionistic fuzzy recommender system. Authors in [17–19] used a fuzzified hybrid model by applying fuzzy logic into user profile features and computed the similarity among fuzzy profiles with a fuzzy distance formula. A context-aware recommender system (CARS) provides additional information for recommending a product. It provides recommendations for those applications which do not support the third attribute apart from users and items, such as recommending a tour package (item) to Tom (user) in winter (context). Here, it is also important to consider the condition (circumstances) for which the recommendations needed. CARS is the hottest topic of research nowadays, and researchers are trying to improve the accuracy and efficiency of the traditional recommender systems by incorporating the contextual information [19]. In literature, for different types of contextual information, different strategies have been used [24, 25]. As discussed in the previous section, context pre-filtering approach was developed by using the item splitting method [26], whereas Baltrunas and Ricci [27] tried to show the experimental comparison between pre- and post-filtering approaches and discussed the dominance of each other in different conditions. Context modeling approach has used the support vector machine classifier for generating contextual-dependent recommendations [28]. Our work in this paper falls into the category of context modeling approach.

In real-world situation, each user tends to choose items based on her personal choice and sometimes users do not give importance to some of the recommended items. In order to improve user's personal recommendation experience, we have to provide such recommendations which user actually likes because in many cases some users like a particular feature of an item but others do not. For example, some user likes action movie, whereas some likes comedy movies. Therefore, we try here the genetic algorithm concept to get the actual user interest to particular feature rather giving the same importance to all features. People are trying to incorporate evolutionary algorithms in the RSs in order to deal with the current issues associated with them. Genetic algorithm [29, 30], particle swarm optimization (PSO) [4, 31], Invenire [32], and invasive weed optimization [33] algorithms have been incorporated in RSs for better recommendations. Ujjin and Bentley [34] presented a genetic algorithm model for refining CF methods before using them in making predictions and showed the performance improvement in the presented approach. Similarly, authors in [17, 18] used the GA and PSO methods in order to learn the optimal priorities for individual features of different users and generated more personalized and accurate movie recommendations.

3 Proposed Genetic Framework

The proposed framework is designed by incorporating contextual information into fuzzy collaborative filtering approach using genetic algorithm. In addition, the framework acquires additional background information including user demographic data and movie genres information in order to form a user model. The framework is partitioned into three phases similar to traditional CF, as shown in Fig. 1. In the following subsections, we will explain the creation of a user profile based on hybrid features, the process of similarity computation by employing the genetic algorithm for each feature and the complete recommendation strategy.

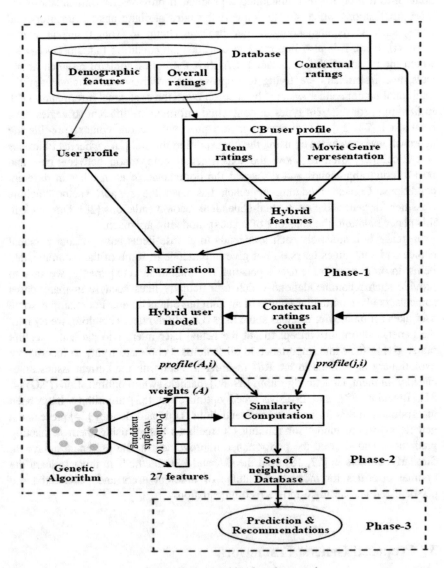

Fig. 1 Framework of the proposed genetic algorithm based approach

3.1 Phase 1—User Profile Generator

The user and movie data is processed into separate user profiles to build a hybrid user model for each user u. The user profile contains the movie rating, demographic information, genre belongingness, and contextual data, as shown in Table 1. The contextual rating count method will be discussed at the end of this subsection.

This user profile model shows 21 different genres where a movie can belong to more than one genre. Here, 1 represents that movie belongs to that particular genre and 0 for otherwise. To know how much a user u_a likes a particular genre G_j, we used the following formula called genre interestingness measure (GIM) [17]. Here, GIM is computed under the guidance of user-provided ratings to movies.

$$GIM(a,j) = \frac{2 \times nf \times RGR(a,j) \times MRGF(a,j)}{RGR(a,j) + MRGF(a,j)}, \tag{1}$$

where MRGF is the modified relative genre frequency for genre G_j to user u_a and computed as

$$MRGF(a,j) = \frac{\sum_{g \in G_j \subset C_i} \delta_3(r_{a,g}) + 2 \times \delta_4(r_{a,g}) + 3 \times \delta_5(r_{a,g})}{3 \times TF(a)}, \tag{2}$$

where δ_n denotes the number of n ratings given by the user. Relative Genre Rating (RGR) is computed as the ratio of u_a's ratings for high-rated items (items which have ratings greater than 2) of G_j to her total given ratings (TR).

$$RGR(a,j) = \frac{\sum_{g \in G_j \subset C_i \geq 3} r_{a,g}}{TR(a)}. \tag{3}$$

Normalization factor (nf) can be taken as the maximum system supported rating (5 in our case) or global average rating $(TR(i)/TF(i))$. Total frequency (TF) represents the total number of items rated by the user. Fuzzy sets for the age and GIM features are shown in Figs. 2 and 3, respectively.

These features are fuzzified to handle the uncertainty problem associated with them. For example, the distance between two users with ages 16 and 19 is 3, while both are from same age category, i.e., teenager. In the same way, in real-life scenario, if we ask someone about her opinion for a movie, then she would say "very bad comedy" or "Excellent action movie"; therefore, we use fuzzy logic for these features to deal with vague concepts and to get as close as possible neighborhood set for the target user. Age feature is treated as a fuzzy variable and

Table 1 A simple user profile model

Rating	Age	Gender	Genre belongingness	Contextual rating Count			
4	31	1	111000010010001100010	$Cf1$	$Cf2$	$Cf3$	$Cf4$

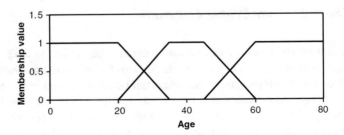

Fig. 2 Age feature fuzzy sets

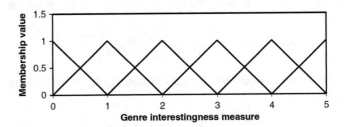

Fig. 3 GIM feature fuzzy sets

classified into three fuzzy sets, namely young, middle, and old. Figure 3 shows six fuzzy sets for GIM feature, namely very bad (VBad), bad, average (Avg), good, very good (VGood), and excellent (Exl).

Table 2 shows the fuzzified values of age and GIM features for a user with age 31 and GIM 2.80.

3.1.1 Contextual Rating Count Method

Out of 12 types of different contextual variables, we extracted only those contextual variables which usually affect our daily life decision-making process. Each context variable is associated with multiple attributes, for instance, season context comprises winter, spring, summer, and autumn as its four attributes. A simple context variable is defined by $Cv[at_1, at_2, at_3, \ldots, at_m]$, where m represents the cardinality of attributes of context Cv. Selected four contexts are listed below:

Table 2 Age and GIM features fuzzified values

	User age			Genre interestingness measure (GIM)					
Quantifiers	Young	Middle	Old	VBad	Bad	Avg	Good	VGood	Exl
Membership value	0.27	0.73	0	0	0	0.2	0.8	0	0

- C1-Time [*morning, afternoon, evening, night*].
- C2-Social [*Alone, Friends, Parents, partner, Colleagues, Public, family*].
- C3-Emotion [*Sad, Neutral, Scared, Surprised, Happy, Angry, Disgusted*].
- C4-Day type [*Weekend, Working day, Holiday*].

Now suppose that a user U_1 has rated a set of movies $(I_1, I_2, \ldots, I_6)$ with above-mentioned four contexts. Then, a sample dataset for the same can be seen in Table 3.

If the user has rated s movies based on x contexts, where each context has associated with y attributes, then the contextual rating count for each context can be computed using the following formula [19]:

$$Cf(x, y) = \frac{\sum_{i \in s}^{|s|} f_{i,y}}{|s|},$$ (4)

where $Cf(x, y)$ shows the normalized rating count for the attribute y of context x and $f_{i,y}$ is the rating count of attribute y in the movie i. Finally, Table 4 shows the normalized rating count of each attribute by dividing the count value with the total number of movies rated by the user U_1.

3.2 Phase 2—Neighborhood Set Selection

In order to generate a neighborhood set for the target user, the main step is similarity computation. In this phase, we use Euclidean global fuzzy distance formula for similarity computation:

$$Gfd(U, V) = \sqrt{\sum_{i=1}^{23} (Lfd(u_i, v_i))^2},$$ (5)

For similarity computation among the fuzzified features, we used the following local fuzzy distance formula:

Table 3 User-rated movies with contexts

User	Movie	C1-time	C2-social	C3-emotion	C4-day type
U_1	I_1	Evening	Family	Neutral	Holiday
U_1	I_2	Afternoon	Family	Happy	Working
U_1	I_3	Night	Colleagues	Surprised	Weekend
U_1	I_4	Morning	Friends	Surprised	Holiday
U_1	I_5	Evening	Colleagues	Happy	Weekend
U_1	I_6	Morning	Alone	Sad	Holiday

Table 4 Normalized rating count value for context attributes

C1-time	Morning	Afternoon	Evening	Night
	2/6 = 0.33	1/6 = 0.16	2/6 = 0.33	1/6 = 0.16
C2-social	Alone	Partner	Friends	Colleagues
	1/6 = 0.16	0/6 = 0	1/6 = 0.16	2/6 = 0.33
	Parents	Public	Family	
	0/6 = 0	0/6 = 0	2/6=0.33	
C3-emotion	Sad	Happy	Scared	Surprised
	1/6 = 0.16	2/6 = 0.33	0/6 = 0	2/6 = 0.33
	Angry	Disgusted	Neutral	
	0/6 = 0	0/6 = 0	1/6 = 0.16	
C4-day type	Holiday	Weekend	Working day	
	3/6 = 0.5	2/6 = 0.33	1/6 = 0.16	

$$\text{Lfd}(u_i, v_i) = \text{dis}(\boldsymbol{u}_i, \boldsymbol{v}_i) \times d(u_i, v_i), \tag{6}$$

where $d(u_i, v_i)$ computes the difference between vectors $\boldsymbol{u}$ and $\boldsymbol{v}$ of size k, and $\text{dis}(\boldsymbol{u}_i, \boldsymbol{v}_i)$ is any vector distance metric which we computed using the following equation in our experiments:

$$\text{dis}(\mathbf{u}_i, \mathbf{v}_i) = \sqrt{\sum_{i=1}^{k} (u_{i,j} - v_{i,j})^2}, \tag{7}$$

where $u_{i,j}$ is the membership value of the ith feature in the jth fuzzy set. This measure computes the distance between features at crisp value level and fuzzy sets level and then combines them using multiplication operator. The gender feature can be selected as fuzzy points with a membership value of 1 if it is the same match and 0 for otherwise. As we can see that user profile contains both non-fuzzy feature (Context count) and fuzzified features (Age, GIM), therefore, we need a complete similarity measure which can be applied to both types of features. To handle this problem, we have combined Eq. (5) with Eq. (7) for effective distance measure between users. Furthermore, we assigned weights w to all features in order to capture the prominence of each feature into users' real-life situation. The updated equation is shown below:

$$\text{Gfd}(\boldsymbol{U}, \boldsymbol{V}) = \sqrt{\sum_{i=1}^{23} w_i \times (\text{Lfd}(u_i, v_i))^2 + \sum_{j=1}^{4} w_j \times \text{dis}(u_j, v_j)}, \tag{8}$$

where w_i represents the weight for the ith feature (age, gender, and 21 movie genres) and w_j denotes the weight for jth feature (four contexts). We use genetic algorithm to learn actual values of these weights for each feature. After computing the distances among users effectively, top 20 users are selected as the neighborhood set for the target user.

3.3 Phase 3—Predictions and Recommendations

After getting the neighborhood set in phase 2, now our task is to predict target user's unseen movie ratings. The main difference between phase 2 and 3 is that phase 2 works only in training set while phase 3 predicts unknown ratings in the test set. Finally, prediction of the target user's unseen items is computed by using the following equation:

$$\text{pre}_{u,i} = \bar{r}_u + m \sum_{u' \in z} \text{Gfd}(u, u') \times (r_{u',i} - \bar{r}_{u'}), \tag{9}$$

where z is the set of neighbors to the target user and m is a normalization factor which is computed as $m = 1/\sum_{u' \in C} |\text{Gfd}(u, u')|$, $\bar{r}_{u'}$ is the user u' average rating.

4 Experiments and Results

In this section, we will discuss about the dataset and setup, parameters used for the experiments, and the evaluation metrics used for the system evaluation.

4.1 Dataset and Setup

We evaluate our proposed approach on LDOS-CoMoDa [35] dataset for the movie domain, which is collected through a survey where users were asked to rate the movies with respect to their particular situation (context). Originally, our dataset contains 2296 users' movie ratings on a scale of 1–5 given by different 121 users on 1232 movies with 12 different contextual attributes (*time, daytype, season, location, weather, social, end emotion, dominant emotion, mood, physical, decision, and interaction*). In order to get relatively significant user ratings (for training dataset), we perform a preprocessing on the complete dataset and extracted only those users who have at least 5 movie ratings. Finally, 48 users with 1964 ratings on 1144 movies have been selected. Similarly, integration of all context attributes makes the system more complex and time-consuming; therefore, out of these 12 contextual attributes, we selected only four important different attributes for fast user profile building. Tenfold cross-validation is used to generate 10 random splits (S1, S2,..., S10) for training and testing sets to deal with the biasness of the system. Each split contains 5 target users and rest of the 43 users are used for similarity computation. The complete ratings of each target user are divided into training and testing sets, where training set (66% ratings) is used to compute similarity among users and neighborhood set formation and testing set (34% ratings) is used for rating prediction and recommendation quality evaluation.

Genetic algorithm is used for evolving profile features' weights by taking population size as 10 for all the experiments, while 30 iterations have been selected as the maximum number of generations for which algorithm runs, each target user is supposed to run 10 times from which the best one is selected at the end of the process. In each generation, 10 individuals are generated and individual which has best fitness value will replace the old individuals. Single-point crossover is applied to the eight randomly selected parents and generates eight new individuals from them. The mutations operation is applied to the rest of two individuals. The user profile contains 27 different features, each of which has 8 bits assigned by an unsigned binary encoding scheme. Decimal value for each feature can be obtained by converting the alleles of the binary genes to decimal. For each feature, the weighting value is computed by dividing each feature's decimal value with the total sum of all features decimal values. The optimal weights of these features can be learned through a supervised learning method. After getting the final set of optimal weights, prediction can be made using the test rating set. Finally, the fitness score of each individual is computed with the help of fitness function shown in the following equation [17]:

$$\text{fitness} = \frac{1}{t_R} \sum_{l=0}^{t_R} |r_l - \text{pre}_l|, \tag{10}$$

where t_R is the training set cardinality, r_l is the actual user rating, and pre_l is the predicted rating for an item l of target user in the training set.

4.2 Performance Measures

To analyze the effectiveness of our proposed approach, we measure its efficiency using two evolution matrices, namely the total coverage of system and mean absolute error (MAE). The average absolute difference between the actual rating $r_{i,j}$ and predicted rating $\text{pre}_{i,j}$ is known as mean absolute error, where minimum MAE indicates better performance of the method:

$$\text{MAE}(i) = \frac{1}{t_i} \sum_{j=1}^{t_i} |pre_{i,j} - r_{i,j}|, \tag{11}$$

Coverage is used to measure the percentage of products for which a system is able to generate predictions.

$$\text{Coverage} = \frac{\sum_{i=1}^{T_n} q_i}{\sum_{i=1}^{T_n} t_i}, \tag{12}$$

where q_i and t_i represent the cardinality of predicted items and cardinality of test ratings of user u_i, respectively.

4.3 Results and Analysis

To validate the performance of our proposed framework, we pick some state-of-the-art collaborative filtering approaches, including the Pearson

Table 5 MAE for 10 splits of four different recommendation approaches

Method/split	MAE value for 10 splits										Average MAE
	S1	S2	S3	S4	S5	S6	S7	S8	S9	S10	
PCF	0.844	0.908	0.984	0.994	1.481	1.463	0.652	0.834	1.309	1.210	10.679
FCF	0.565	0.534	0.785	0.630	1.305	1.354	0.642	0.657	0.963	1.110	8.545
CA-FCF	0.528	0.496	0.763	0.536	1.232	1.149	0.623	0.581	0.914	0.956	7.778
GA-CA-FCF	0.507	0.462	0.74	0.519	1.186	1.084	0.603	0.573	0.813	0.887	**7.374**

Table 6 Coverage for 10 splits of four different recommendation approaches

Method/split	Coverage value for 10 splits										Average coverage
	S1	S2	S3	S4	S5	S6	S7	S8	S9	S10	
PCF	0.357	0.381	0.31	0.528	0.521	0.312	0.428	0.443	0.341	0.479	0.41
FCF	0.379	0.406	0.311	0.532	0.507	0.25	0.428	0.446	0.341	0.39	0.399
CA-FCF	0.456	0.451	0.32	0.54	0.536	0.32	0.457	0.661	0.365	0.455	0.4561
GA-CA-FCF	0.489	0.479	0.32	0.548	0.556	0.329	0.461	0.661	0.405	0.478	**0.4726**

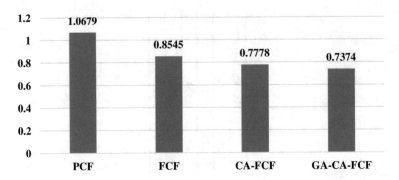

Fig. 4 Comparison of average MAE for four different recommendation approaches

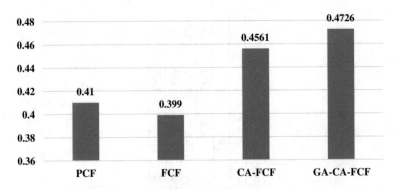

Fig. 5 Comparison of average coverage for four different recommendation approaches

collaborative filtering (PCF), fuzzy collaborative filtering (FCF) [17], and context-aware fuzzy CF (CA-FCF) [19]. The MAE and coverage of target users for all 10 splits for the examined approaches are presented in Tables 5 and 6. Results show that GA-CA-FCF outperforms all three examined approaches PCF, FCF, and CA-FCF in terms of MAE for all 10 splits. FCF, CA-FCF, and GA-CA-FCF use the improved collaborative filtering technique by using additional user–item information and fuzzy logic.

We can see from Fig. 4 that complete MAE of GA-based approach GA-CA-FCF is always lesser than the other traditional approaches, where lower MAE value indicates better performance of the proposed method. Average system coverage obtained from all discussed approaches is graphically represented in Fig. 5. The higher coverage by GA-CA-FCF illustrates that the accuracy gets enhanced by obtaining a large number of predictions as compared to rest of the approaches.

The results presented above reveal that our proposed approach has superior performance in terms of both MAE and coverage over movie dataset. We can conclude from the above-presented experimental results that outcome of all previous methods measured by different metrics shows lower performance than the genetic algorithm-based approach.

5 Conclusion and Future Work

In this work, we employed genetic algorithm in recommender system to compute efficient similarity among different users, based on the comparison between several user profile features, for the movie recommendations system. Our main focus is to improve the similarity computation among the user profile and generate the optimal neighborhood set of traditional collaborative filtering approach, and to improve the performance of CARSs. To achieve this, we derive a novel genetic framework which generates recommendations using three different phases similar to the traditional CF technique. In the first phase, we propose to build different users contextual profiles for all contextual conditions. To get actual user interest, we further use fuzzy logic on age and GIM features of the user model. In the second phase, we compute the similarity among users by employing GA and generating recommendations in the third phase of the framework. We have performed experiments on a real-world movie dataset and have compared with the traditional collaborative filtering approaches with respect to their prediction accuracy. Experimental results show that use of the genetic algorithm is a promising way to improve the effectiveness of collaborative filtering-based recommender system. In future, there are a lot of other challenges that need to be explored; for example, our work in this paper is limited to movie domain and it would be interesting to extend this work in other domains also (e.g., music and books). Furthermore, we also plan to employ fuzzy logic and statistical methods for contextual features to deal with the uncertainty issue associated with them.

References

1. Bobadilla J, Ortega F, Hernando Antonio, Gutiérrez Abraham (2013) Recommender systems survey. Knowl Based Syst 46:109–132
2. Resnick P, Iacovou N, Suchak M, Bergstrom P, Riedl J (1994) GroupLens: an open architecture for collaborative filtering of netnews. In: Proceedings of the 1994 ACM conference on computer supported cooperative work, pp 175–186
3. Aggarwal CC (2016) Neighborhood-based collaborative filtering. In: Recommender systems. Springer International Publishing, New York, pp 29–70
4. Klir G, Yuan B (1995) Fuzzy sets and fuzzy logic, vol 4. Prentice hall, New Jersey
5. Abbas A, Zhang L, Khan SU (2015) A survey on context-aware recommender systems based on computational intelligence techniques. Computing 97(7):667–690
6. Adomavicius G, Ramesh S, Sen S, Alexander T (2005) Incorporating contextual information in recommender systems using a multidimensional approach. ACM Trans Inf Syst 23(1): 103–145

7. Adomavicius G, Tuzhilin A (2011) Context-aware recommender systems. Springer, US, In Recommender systems handbook, pp 217–253
8. Golberg DE (1989) Genetic algorithms in search, optimization, and machine learning. Addison-Wesley, Boston
9. Aggarwal, Charu C (2016) Content-based recommender systems. In: Recommender Systems, Springer International Publishing, New york, pp 139–166
10. Bell RM, Koren Y (2007) Improved neighborhood-based collaborative filtering. In: KDD cup and workshop at the 13th ACM SIGKDD international conference on knowledge discovery and data mining, pp 7–14
11. Salakhutdinov R, Mnih A, Hinton G (2007) Restricted Boltzmann machines for collaborative filtering. In: Proceedings of the 24th international conference on Machine learning, ACM, pp 791–798
12. Rennie JDM, Srebro N (2005) Fast maximum margin matrix factorization for collaborative prediction. In: Proceedings of the 22nd internationl conference on Machine learning, ACM, pp 713–719
13. Salakhutdinov R, Mnih A (2008) Bayesian probabilistic matrix factorization using Markov chain Monte Carlo. In: Proceedings of the 25th international conference on machine learning, ACM, pp 880–887
14. Sarwar B, Karypis G, Konstan J, Riedl J (2001) Item-based collaborative filtering recommendation algorithms. In:Proceedings of the 10th international conference on World Wide Web, ACM, pp 285–295
15. Linden G, Smith B, York J (2003) Amazon.com recommendations: Item-to-item collaborative filtering. IEEE Internet Comput 7(1):76–80
16. Liu J, Dolan P, Pedersen ER (2010) Personalized news recommendation based on click behavior. In: Proceedings of the 15th international conference on intelligent user interfaces, ACM, pp 31–40
17. Al-Shamri MYH, Bharadwaj KK (2008) Fuzzy-genetic approach to recommender systems based on a novel hybrid user model. Expert Syst Appl 35(3):1386–1399
18. Wasid M, Kant V (2015) A particle swarm approach to collaborative filtering based recommender systems through fuzzy features. Procedia Comput Sci 54:440–448
19. Wasid M, Kant V, Ali R (2016) Frequency-based similarity measure for context-aware recommender systems. In: International conference on advances in computing, communications and informatics, IEEE, pp 627–632
20. Melville P, Mooney RJ, Nagarajan R (2002) Content-boosted collaborative filtering for improved recommendations. In: *Aaai/iaai*, pp 187–192
21. Gunawardana A, Meek C (2008) Tied boltzmann machines for cold start recommendations. In: Proceedings of the 2008 ACM conference on recommender systems, ACM, pp 19–26
22. Tamayo LFT (2014) Fuzzy recommender system. In: Smart participation, Springer International Publishing, New york, pp 47–81
23. Thong, NT (2015) Intuitionistic fuzzy recommender systems: an effective tool for medical diagnosis. Knowl Based Syst, 74:133–150
24. Liu X, Aberer K (2013) SoCo: a social network aided context-aware recommender system. In: Proceedings of the 22nd international conference on World Wide Web, ACM, pp 781–802
25. Yin H, Cui B, Chen L, Hu Z, Huang Z (2014) A temporal context-aware model for user behavior modeling in social media systems. In: Proceedings of the 2014 ACM SIGMOD international conference on management of data, ACM, pp 1543–1554
26. Baltrunas L, Ricci F (2009) Context-based splitting of item ratings in collaborative filtering. In: Proceedings of the third ACM conference on recommender systems, ACM, pp 245–248
27. Baltrunas L, Ricci F (2014) Experimental evaluation of context-dependent collaborative filtering using item splitting. User Model User-Adap Inter 24(1–2):7–34
28. Oku K, Nakajima S, Miyazaki J, Uemura S (2006) Context-aware SVM for context-dependent information recommendation. In: Proceedings of the 7th international conference on mobile data management, IEEE computer society, p 109

29. Tyagi S, Bharadwaj KK (2013) Enhancing collaborative filtering recommendations by utilizing multi-objective particle swarm optimization embedded association rule mining. Swarm Evol Comput 13:1–12
30. Ar Y, Bostanci E (2016) A genetic algorithm solution to the collaborative filtering problem. Expert Syst Appl 61:122–128
31. Ujjin S, Bentley PJ (2002) Learning user preferences using evolution. In: Proceedings of the 4th Asia-Pacific conference on simulated evolution and learning, Singapore
32. da Silva, EQ, Camilo-Junior CG, Pascoal LML, Rosa TC (2016) An evolutionary approach for combining results of recommender systems techniques based on collaborative filtering. Expert Syst Appl 53:204–218
33. Rad HS, Lucas C (2007) A recommender system based on invasive weed optimization algorithm. In: 2007 IEEE Congress on Evolutionary Computation, IEEE, pp 4297–4304
34. Ujjin S, Bentley PJ (2003) Particle swarm optimization recommender system. In: Proceedings of the IEEE Swarm Intelligence Symposium, SIS'03, pp 124–131
35. Odic A, Tkalcic M, Tasic JF, Košir A (2012) Relevant context in a movie recommender system: users' opinion versus statistical detection. ACM RecSys 12

Enhanced Multi-criteria Recommender System Based on AHP

Manish Jaiswal, Pragya Dwivedi and Tanveer J. Siddiqui

1 Introduction

Today, Internet and web technologies have a find a place in our daily life. An individual who is not much aware of technology is also using it in the day-to-day decision-making process. For example, in order to decide which gadget to buy, which book to read, which movie to watch, which college to choose for admission? A huge amount of data is being uploaded every day on the Internet. This includes informative content as well as data in the form of reviews, recommendations, ratings, suggestions, etc. leading to situation commonly known as "information overload". People seeking help from the Internet in decision-making find it difficult to cope with this huge amount of information without automatic aids. Walker realized this situation in early 90s and pointed out that information overload leads to a situation where the input to one's decision-making process exceeds the "capacity to assimilate and act on the information as well as the ability to evaluate every alternative" [1]. Since then the amount of digital content has increased manifold. This results in an increased demand for automated tools and techniques to combat with the problem of "information overload". Research community responded to this problem through the development of information filtering systems to provide personalized solutions to users. More and more research is focusing on the development of personalization and recommendation techniques. A recommender system

M. Jaiswal (✉) · T. J. Siddiqui
Department of Electronics & Communication, University of Allahabad, Allahabad, India
e-mail: manish.jk50@gmail.com

T. J. Siddiqui
e-mail: siddiqui.tanveer@gmail.com

P. Dwivedi
Department of Computer Science Engineering Department, MNNIT, Allahabad, India
e-mail: pragya.dwijnu@gmail.com

© Springer Nature Singapore Pte Ltd. 2017
R. Ali and M. M. S. Beg (eds.), *Applications of Soft Computing for the Web*,
https://doi.org/10.1007/978-981-10-7098-3_3

(RS) is a tool, which helps to overcome the problem of information overload by providing recommendations based on the individual's interests.

RS have found useful applications in many diversified areas, such as in education, social- media, financial services, question–answering system, agriculture, health care, and so on [2, 3]. A number of personalized recommender systems are already in place [4, 5]. These systems offer great opportunities in several domains including business, e-commerce, e-learning, web services, e-government, and e-tourism. More and more successful RS are being developed for real-world application [3].

In this chapter, we discuss the basic concepts involved in the development of recommender systems and propose a method for multi-criteria decision-making based on the analytic hierarchy process (AHP) for college selection. With the increasing competition among colleges, the process of selecting a college that meets most of the desired objective of a student has become a difficult task. There is a number of factors that can affect the student's choice, such as fee, quality of education, social activity, faculty, location, availability of hostels. The decision-making process has to deal with multiple and often conflicting criteria. AHP has been widely used in decision making in the presence of a mix of qualitative and quantitative, and sometimes conflicting factors [6]. This makes AHP an obvious choice for college selection process. The rest of the chapter is organized as follows.

Section 2 presents the basics of a recommender system. In Sect. 3, we discuss various recommendation approaches. Section 4 introduces multi-criteria recommender systems and briefly reviews existing work. Section 5 presents experimental details. Finally, we conclude in Sect. 6.

2 Recommender System Basics

A Recommender System (RS) is defined as "any system that produces individualized recommendations as output or has the effect of guiding the user in a personalized way to interesting or useful objects in a large space of possible options" [7]. RSs help us in combating the information explosion problem by providing a personalized recommendation in a variety of application areas, such as in the movie, news filtering, electronic consumer, education, social tagging, image, and in the mobile recommendation. For example, in the entertainment area, Netflix[1] displays the predicted ratings of a movie in order to help the user decide which movie to rent. Online shopping websites provide ratings of various items which a promising customer may use in deciding which item to purchase. When a user purchases an item these sites recommend a list of items which have been purchased by other users who have purchased this item.

---

[1]https://www.netflix.com/.

Table 1 User-item rating matrix

u/i	i_1	i_2	i_3	i_4	i_5
u_1	4	3	5	3	2
u_2	4	2	1	4	1
u_3	4	1	4	3	2
u_4	4	2	2	3	2

The recommendation problem can be viewed as a rating estimation problem. RSs have to predict the rating of unseen items based on the rating provided by the user for the seen items. After estimation, RS can recommend items with the highest rating to the target user.

Let U be the set of users, I be the set of all possible items and R be the set of ratings that users give to an item. The rating value is usually nonnegative integer value within a certain range assigned to an item by a user. For each user u in U, a recommender system attempts to find an item i from I that maximizes the rating estimation:

$$\forall u \in U, i' = \arg \max_{i \in I} R_{u,i}, \tag{1}$$

where, $R_{u,i}$ is rating provided by user u on item i.

Table 1 represents sample user-item rating matrix for four users ($u1, u2, u3, u4$) and 5 items ($i1, i2, i3, i4, i5$).

The rating r can be considered a utility function that measures the usefulness of an item i for a user u:

$$r : U \times I \to R \tag{2}$$

3 Recommendation Approaches

RSs are broadly classified into three categories based on their approach to recommendation: collaborative filtering approach, content-based approach, and hybrid approaches [8].

3.1 Collaborative Filtering

Collaborative Filtering (CF) approach uses likings of a group of other users having similar taste in order to make a recommendation for an active user. The similarity between two users is calculated based on the similarity in their rating history. The concept of collaborative filtering is depicted in Fig. 1. The user B prefers an item and user A has similar taste to user B so, the same item would be recommended to

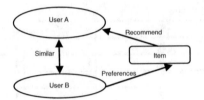

Fig. 1 Collaborative filtering approach

user A. Schafer et al. [9] referred this type of CF as "people-to-people collaborative filtering." There is another approach to CF referred in the literature as "item-to-item collaborative filtering." Item-to item CF provides a recommendation by comparing similar items purchased or rated by a user and combines those items in a recommendation list.

CF is the most popular and widely implemented technique in recommender system [7]. However, it suffers from the following limitations [4]:

1. New user problem—One of the problems with CF is that it requires enough ratings of a user, to begin with. For a user which is new to the system, no rating data is available. So, CF approach will find it difficult to recommend him correctly.
2. New-item problem—Similar problem arises for items which have been newly added to the system. Such items lack users' rating and hence, these items will not be recommended by CF-based recommender system. This problem can be addressed by using the hybrid approach described in Sect. 3.3.
3. Sparsity—Quite often, the number of available ratings is very small. For example, there may be a movie which is rated by few number of users. Collaborative filtering faces such problem to predict rating of such movies. The RS are required to include an effective mechanism to cope with the data sparseness problem.

3.2 Content-Based Recommendation

In content-based recommendation (CBR), the system compares the content of items from user's previous likings and recommends those which are most similar to user's profile. For example, if a user has positively rated a book that belongs to cuisine genre, then the system can learn to recommend other books from this genre. In Fig. 2, there are two items X and Y. Let a user prefers item X while item Y having same features similar to item X, then item Y will also be recommended to the user. Here, we need to have a measure of similarity between items which is calculated based on their feature description.

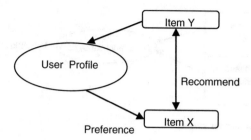

Fig. 2 Content-based recommendation

CBR has several limitations which are discussed below:

1. Limited content analysis—CBR is suitable for text-based applications because computers can parse it automatically. For images, audio, video, etc. automatic parsing is not an easy task and most of the recommender system use manually assigned features for recommending multimedia content. One more problem is that, if two items have same features we cannot differentiate between them.
2. Overspecialization—CBR system maintains a user profile and uses it in the recommendation. If a user has a very high rating for an item, it may happen that other items are never recommended to him/her. This problem can be addressed by adding some randomness (genetic algorithm).
3. New user problem—The user which is newly added to the system has no previous likings, and hence, CBR system will fail to recommend him correctly [8].

3.3 Hybrid Recommender System

Hybrid RSs are based on the combination of the content-based filtering and collaborative filtering techniques. A hybrid RS combines both techniques and tries to use the advantages of one technique to fix the disadvantages of other technique [10]. For instance, Collaborative filtering approach suffers from new-item problems, i.e., they cannot recommend items that have no ratings. This problem is not occurred in content-based approaches since it uses item description to recommend new items which are typically available (Fig. 3).

Hybrid recommender system incorporates various hybridization techniques, such as feature augmentation, feature combination, weighting, cascading [2], and so on [11]. The main objective of a hybrid RS is to provide better recommendations which enhance the overall performance of the system with higher accuracy.

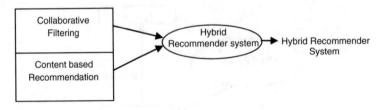

Fig. 3 Hybrid recommender system

4 Multi-criteria Recommender System (MCRS)

So far, we talked about a decision-making situation where a user assigns a single rating to an item. In recent work, this assumption has been considered as limited [8, 12] because the suitability of the recommended item for a particular user may depend on several criteria that the user takes into consideration when making the choice. The involvement of multiple criteria makes it difficult for a user to choose the best alternative from a set of available alternatives. A RS can use this additional information provided by multi-criteria ratings to provide more accurate recommendations to its user. MCRS extends the capabilities of a general RS by considering ratings as multi-valued instead of a single value. The utility-based formulation of the multi-criteria recommendation problem can be represented as follows:

$$r : U \times I \rightarrow R_1 \times R_2 \times R_3 \times \cdots R_m \tag{3}$$

where R_i represents the possible rating values for each individual criterion of item i typically on some numeric scale.

In a multi-criteria recommender system, the rating of an item is based upon several parameters. In recent years, multi-criteria recommender systems have become enormously common in a variety of applications [4] such as in entertainment, business, hotel industry, e-commerce, education, consumer goods, health care, and so on. The online shopping website buy.com recommends the best option to users by using multi-criteria ratings for consumer electronics (e.g., price, battery life, performance, and display). Online restaurant guide, Zagat's Guide recommends the best restaurant based on three criteria (e.g., decor, food, and service).

Decision-making is the process of identifying a decision and choosing alternatives based on the preferences of the decision maker. Making a decision implies that there are alternative choices to be considered, and in such a case, we want not only to identify as many of these alternatives as possible but to choose the one that best fits with our goals, objectives, desires, values, and so on. MCDM refers to making decisions in the presence of multiple criteria. In our day-to-day life, everything we do consciously or unconsciously is the result of some decision.

The objective of multi-criteria decision analysis is to assist a decision maker in choosing the best alternative when multiple criteria conflict and compete with each other. There is no unique optimal solution for these problems, typically, it is necessary to use decision maker's desire to differentiate between solutions.

Basically, MCDM problem is based on whether the solutions are explicitly or implicitly defined. There are various methods of MCDM, such as SAW (Simple Additive Weighting), AHP (Analytic Hierarchy Process), ELECTRE (Outranking), Multi-attribute utility theory (MAUT), VIKOR, and TOPSIS (Technique for Preference by Similarity to the Ideal Solution). Here, in this chapter, we report AHP-based MCDM technique. In this section, we briefly introduce AHP and review earlier works done using AHP.

4.1 AHP (Analytic Hierarchy Process)

The AHP is an excellent MCDM tool for dealing with the complex decision-making problems, and assists the decision maker to set priorities for making the best decision [6]. AHP decision-making approach involves structuring multiple choice criteria into a hierarchy, assessing the relative importance of these criteria, comparing alternatives for each criterion and determining an overall ranking of the alternatives. AHP is used in various decision-making situations, such as in stock market, education, management, economic forecasting, business, resource allocation, and many more. The first step of AHP is to decompose the problem into a hierarchy of subproblems, by arranging the relevant factors of the problem into a hierarchical structure that descends from an overall goal to criteria, sub-criteria and alternatives, in successive levels. The second step of AHP comprises the elicitation of pair-wise comparison judgments from the decision making body. Here, the decision maker is asked to assess the relative importance of criteria with respect to the goal, through pair-wise comparisons. The output of this preference elicitation process is a set of verbal answers of the decision maker, which are subsequently codified into a nine-point intensity scale. The standard preference scale used in the AHP method was proposed by Saaty [6] shown in Table 2.

After the determination of the pair-wise comparisons among all the criteria, AHP converts the corresponding decision maker's evaluation into vector of priorities by finding the first eigenvector of the criteria matrix. This vector has information about the relative priority of each criterion with respect to the global goal.

Afterward, a weighting and summing step yields the final result of AHP, which are the orderings of the alternatives based on a global indicator of priority. The alternative with the largest value of this global score is the most preferred one.

The basic idea behind AHP is to convert subjective assessments of relative importance into a set of overall scores and weights. The assessments are subjective since, they reflect the perception of the decision maker, and are based on pair-wise comparisons of criteria/alternatives.

Table 2 Standard preference scale [6]

Preference level	Numerical value
Equally preferred	1
Equally to moderately preferred	2
Moderately preferred	3
Moderately to strong preferred	4
Strongly preferred	5
Strongly to very strongly preferred	6
Very strongly preferred	7
Very strongly to extremely preferred	8
Extremely preferred	9

The steps for implementing the AHP process are listed below [6]:

1. Define the Objectives.
2. Identify the Criteria/Attributes.
3. Choose the Alternatives.
4. Establish the Hierarchy.
5. Construct Pair-wise Comparison matrices.
6. Synthesize Judgments.
7. Calculate Consistency (C.I) Index.
8. Compare Criteria and Alternatives.
9. Calculate Final Rankings.

The Random Consistency Index (RI) can be observed in Table 3 as follows:

It is widely used in many real-life applications include technology choice [13], supplier selection [14], college selection, resource allocation, project selection, and so on.

4.2 Related Work

The last two decades have witnessed an increasing amount of research in the area of RS. Many of these works focus on MCDM problem. The prominent decision-making techniques include SAW (Simple Additive Weighting), AHP (Analytical Hierarchy Process) [15–18], TOPSIS (Technique for Order Preference by Similarity to Ideal Solution) [17, 19, 20] and fuzzy AHP.

AHP method is more suitable when an attribute hierarchy has more than three levels [15]. Frair et al. [21] proposed an AHP-based method for academic

Table 3 Random consistency index

n	1	2	3	4	5	6	7	8	9
RI	0	0	0.58	0.90	1.12	1.24	1.32	1.41	1.45

curriculum evaluation. Their model was based on the responses from the parties (employee, students, faculties, etc.), curriculum components (math, design, science, etc.), and curriculum alternatives.

Kang and Seong [22] proposed a procedure for evaluating the alarm-processing system in a nuclear power plant control room using informational entropy-based measure and AHP.

Cebeci [23] proposed fuzzy AHP-based decision support system for Enterprise Resource Planning (ERP) systems for textile industry by using a balanced scorecard.

Ozcan and Celebi [24] applied AHP, Topsis, and Grey Theory on a warehouse location selection problem involving various criteria, such as unit price, stock holding capacity, average distance to shops, and average distance to main suppliers.

Dougligeris and Pereira [13] proposed the use of analytical hierarchy process in a telecom industry to help customers in choosing a telecommunication company that best specifies their needs.

Yang et al. [25] applied the analytic hierarchy process in location selection for a company. The location decision often depends on the type of business. For the industrial organization, the objective is to minimize the cost while for service organization; the primary concern is to maximize revenue.

Rezaiana [26] applied AHP-based MCDM technique for health-safety and environmental risk of refineries.

SaeedZaeri et al. [27] applied Topsis method for supplier selection problem which is MCDM problem in supply chain management. They considered several criteria, such as financial stability, delivery lead time, accessibility, reliability, transportation cost.

Chang and Chen [28] proposed a fuzzy-based MCDM method for location selection for the distribution center Problem.

Dagdeviren et al. [29] proposed a model to help defense personnel for weapon selection using AHP and fuzzy TOPSIS technique.

5 Proposed Work

The proposed work uses AHP for college selection. There are many criteria which one uses to decide the college where s/he will apply for admission. These include quality of education, fee structure, hostel facility, location, etc. Some of these criteria may conflict each other. We use AHP because it is designed for multi-criteria decision-making problems conflicting goals. The dataset used in this work is collected from USNEWS dataset which contains rating corresponding to various criteria. The proposed method can be applied in any application where ratings are associated with several criteria.

5.1 Overview of the Proposed Method

In the proposed work, we have considered eight colleges (Brown, Columbia, Cornell, Dartmouth, Harvard, Princeton, U. Penn, and Yale) and eight criteria (Tuition, Acceptance rate, Salary, Education, Social life, Faculty accessibility of faculty, and near city) in order to select the best college. The AHP method is used for Multi-criteria decision-making recommendation (Fig. 4).

The basic idea behind AHP is to convert subjective assessments into a set of overall scores and weights.

It is a method for solving complex decision making based on the alternatives and multiple criteria. It is also a process for developing a numerical score to rank each decision alternative based on how well each alternative meets the decision maker's criteria. The steps involved in the proposed method are:

1. Determine the weight for each criterion.
2. Determine the ratings for each decision alternative for each criterion.
3. Calculate the weighted average rating for each alternative and select the highest scoring one.

   Step 1: Determine the weight for each criterion in a matrix of $n \times n$. where n is the number of criteria. We perform following stages in step 1 for determining the weight-

   (a) Construct a pair-wise comparison matrix as shown in Table 4. Here, each a_{ij} represents the importance of criterion i with respect to criterion j. If $a_{ij} > 1$, then the user gives more importance to criterion i than criterion j. If $a_{ij} = 1$ for $i = j$, then it represents both criterion have the same importance with respect to each other.
   (b) Multiply the values in each row together and calculate the nth root of the product.
   (c) Normalize the nth root of products to get the appropriate weights. Now, we calculate the CR using the formula:

$$CR = \frac{CI}{RI},\qquad(4)$$

where, CI is consistency Index and RI is Random Index. RI value is obtained using Table 3 and CI is calculated as follows:

$$CI = \frac{\lambda_{max} - n}{n - 1},\qquad(5)$$

where, λ_{max} is the largest eigenvalue and n is a total number of criteria.

If the value of CR is less than 0.1 then the pair-wise comparisons are considered consistent and we proceed ahead with step 2 otherwise corrective measures are required.

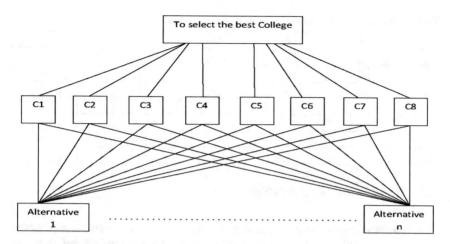

Fig. 4 The proposed method

Table 4 Pair-wise comparison matrix

	C_1	C_2	C_3	...	C_n
C_1	a_{11}	a_{12}	a_{13}	...	a_{1n}
C_2	a_{21}	a_{22}	a_{23}	...	a_{2n}
C_3	a_{31}	a_{32}	a_{33}	...	a_{3n}
.	.	.	.	...	.
.	.	.	.	...	.
.	.	.	.	...	.
C_n	a_{n1}	a_{n2}	a_{n3}	...	a_{nn}

Step 2: Determine the ratings for each decision alternative considering each criterion. This can be obtained in following stages-

(a) First, develop a pair-wise comparison matrix of alternatives for each criterion as shown in Table 5.

(b) Multiply the values in each row together and calculate the nth root. Where $A_1, A_2, \ldots, A_n$ are the alternatives.

(c) Normalizing the nth root of products to get the appropriate weights.

(d) Calculate the consistency ratio (CR) using the formula:

$$CR = \frac{CI}{CR} \qquad (6)$$

Step 3: Then, we calculate the weight value for each alternative which provides the rank of an alternative. Finally, we select the one which has maximum weight value for the recommendation.

Table 5 Pair-wise comparison matrix with respect to criteria

	A_1	A_2	A_3	$\cdots$	A_n
A_1	a_{11}	a_{12}	a_{13}	$\cdots$	a_{1n}
A_2	a_{21}	a_{22}	a_{23}	$\cdots$	a_{2n}
A_3	a_{31}	a_{32}	a_{33}	$\cdots$	a_{3n}
.	.	.	.	$\cdots$	.
.	.	.	.	$\cdots$	.
.	.	.	.	$\cdots$	.
A_n	a_{n1}	a_{n2}	a_{n3}	$\cdots$	a_{nn}

5.2 Experiment and Results

This section provides the experiment and result analysis. For performance evaluation of our result, we have used quality metric precision which refers to the closeness of two or more measurements to each other and Kendall's tau rank correlation coefficient (τ) which measures the strength of the relationship between the preferences of two students.

5.2.1 Dataset Description

We have conducted experiments on a dataset collected from U. S. news dataset. We have extracted ratings for eight different colleges on eight different criteria. The original rating in the resource and the extracted data is presented in the following Table 6.

The criteria used in the evaluation of the proposed method is (1) Tuition, (2) Acceptance, (3) Starting Salary (Salary), (4) Education (Edu), (5) Social Life (Soc Life), (6) Near City (N-city), (7) Going Well, and (8) Faculty Accessibility (Fac-Acc). Here, we can divide each criterion into many sub-criteria, and each sub-criterion into smaller sub-criteria, and so on, until we reach to our alternatives (college options) for each of the smallest sub-criterion. The alternatives would be a list of colleges that best fulfill ones desired criteria. For simplicity, the ratings are A, A−, B+, B, B−, and C+ provided by users are converted into numerical values. Here, A is the maximum and C + is the minimum value. The numerical values of these ratings are shown in the Table 7.

After converting the ratings in numerical values, the next step is to normalize these ratings. The AHP procedure is based on the pair-wise comparison of decision elements with respect to criteria. A pair-wise comparison matrix is formed for number each alternative. A pair-wise comparison matrix for the "near city" criteria using rating averaged is presented in the Table 8. Similarly, we calculate pair-wise comparison for other criteria.

Table 9 shows the weighted value (final score) obtained for five students using their preferences on four criteria: salary, near city, going well, and faculty accessibility.

Table 6 Users rating of various criteria

College/criteria	Tuition	Accept (%)	Salary	Edu	Soc life	N-city	Going well (%)	Fac-acc
Brown (CL1)	$37,637	39.40	$88,000	A−	A−	A−	84.9	A−
Columbia (CL2)	$40,592	17.20	$1,01,428	B+	B	A−	76.9	B+
Cornell (CL3)	$38,800	36.40	$94,370	A−	B	B−	100.0	B+
Dartmouth (CL4)	$40,650	20.30	$1,00,220	A	A−	B−	93.8	A
Harvard (CL5)	$39,600	14.90	$1,08,785	B	B−	B+	52.9	B−
Princeton (CL6)	$40,000	38.20	$90,000	B+	B	C+	80.8	B+
U. Penn (CL7)	$39,650	18.80	$1,05,684	A−	A−	A−	81.0	B+
Yale (CL8)	$39,500	22.00	$92,213	A−	B+	B	68.0	A−

Table 7 Rating values

A	100
A−	83
B+	67
B	50
B−	33
C+	17

We evaluated the performance of obtained result in terms of precision and correlation coefficient (Kendall's tau). The observed average precision for five students is 0.875 which is quite significant.

Kendall's tau correlation coefficient (τ) determines the similarity of the ranking between variables. It assesses the statistical associations based on the rank. It is widely used rank correlation coefficient. The value of Kendall's tau correlation coefficient (τ) specifies how much the rank of alternatives between two students are similar to each other. The Kendall Tau correlation coefficient is defined as (Table 10):

$$\tau = \frac{(\text{number of concordant pairs}) - (\text{number of discordant pairs})}{\frac{1}{2}n(n-1)} \tag{7}$$

Here τ_{ij} represents the correlation between student i and j. The value of τ close to 1 indicates that ranking obtained for different colleges is quite similar. For example, the value of $\tau_{2,4}$ is 0.9285 which is larger than the value of $\tau_{2,5}$. This is because the preferences of student 2 for various colleges match more closely with student 4 than 5. The list of colleges recommended to a student using is same as the list of recommended colleges provided in the dataset website. This validates the experimental results and the use of AHP for the college selection task.

Table 8 Pair-wise comparison matrix of colleges with respect to near city

	CL 1	CL 2	CL 3	CL 4	CL 5	CL 6	CL 7	CL 8
CL 1	1.000000	1.000000	2.534247	2.534247	1.241611	4.868421	1.000000	1.666667
CL 2	1.000000	1.000000	2.534247	2.534247	1.241611	4.868421	1.000000	1.666667
CL 3	0.394595	0.394595	1.000000	1.000000	0.489933	1.921053	0.394595	0.657658
CL 4	0.394595	0.394595	1.000000	1.000000	0.489933	1.921053	0.394595	0.657658
CL 5	0.805405	0.805405	2.041096	2.041096	1.000000	3.921053	0.805405	1.342342
CL 6	0.205405	0.205405	0.520548	0.520548	0.255034	1.000000	0.205405	0.342342
CL 7	1.000000	1.000000	2.534247	2.534247	1.241611	4.868421	1.000000	1.666667
CL 8	0.600000	0.600000	1.520548	1.520548	0.744966	2.921053	0.600000	1.000000

Table 9 Final score of each college

Score	Student 1	Student 2	Student 3	Student 4	Student 5
Brown	0.140857	0.134874	0.138627	0.132938	0.128745
Columbia	0.123939	0.123336	0.125956	0.125313	0.125959
Cornell	0.125176	0.124617	0.125990	0.124646	0.123891
Dartmouth	0.129961	0.132959	0.127628	0.130672	0.130068
Harvard	0.109157	0.112192	0.113368	0.116006	0.121070
Princeton	0.115609	0.115841	0.115197	0.116427	0.115647
U. Penn	0.135154	0.135022	0.134641	0.133847	0.134562
Yale	0.120146	0.121159	0.119594	0.120150	0.120058

Table 10 Value of Kendall's Tau correlation coefficient

$\tau_{1,2}$	0.9285
$\tau_{1,3}$	0.9285
$\tau_{1,4}$	0.8572
$\tau_{1,5}$	0.7143
$\tau_{2,3}$	0.9285
$\tau_{2,4}$	0.9285
$\tau_{2,5}$	0.8572
$\tau_{3,4}$	0.9285
$\tau_{3,5}$	0.8572
$\tau_{4,5}$	0.7857

6 Conclusion and Future Work

In this study, we evaluate an AHP-based system for recommending a list of colleges to students on the basis of their preferences. We observed an average precision of 0.875. The results on a sample dataset demonstrate that AHP can be used effectively in recommendation systems for college selection. The proposed method can be enhanced using fuzzy AHP. The deep learning techniques can also be used to provide more accurate and personalized recommendation in the future.

References

1. Walker JP (1971) Decision-making under conditions of information overload: alternative response modes and their consequences, ERIC Clearinghouse
2. Isinkaye FO, Folajimi YO, Ojokoh BA (2015) Recommendation systems: principles, methods and evaluation. Egypt Inform J 16(3):261–273
3. Jie L, Dianshuang W, Mingsong M, Wei W, Guangquan Z (2015) Recommender system application developments: a survey. Decis Support Syst 74:12–32

4. Adomavicius G, Tuzhilin A (2005) Toward the next generation of recommender systems: a survey of the state-of-the-art and possible extensions. IEEE Trans Knowl Data Eng 17:734–749
5. Manouselis N, Drachsler H, Vuorikari R, Hummel H, Koper R (2009) Recommender systems in technology enhanced learning, Recommender Systems Handbook, pp 387–415
6. Saaty TL (2008) Group decision making: drawing out and reconciling differences, RWS Publications, ISBN: 188-8-603-089
7. Burke R (2002) Hybrid recommender systems: survey and experiments. User Model User-Adapt Interact 12(4):331–370
8. Adomavicius G, Kwon YO (2007) New recommendation techniques for multi-criteria rating systems. IEEE Intell Syst 22(3):1548–1555
9. Schafer JB, Konstan JA, Riedl J (2001) E-commerce recommendation applications. Data Min Knowl Discov 5:115–153
10. Bobadilla J, Ortega F, Hernando A, Gutierrez A (2013) Recommender systems survey. Knowl-Based Syst 46:109–132
11. Jannach D, Zanker M, Felfernig A, Friedrich G (2011) Recommender systems: an introduction, Cambridge University Press, p 336 ISBN: 978-0-521-49336-9
12. Fandel G, Spronk J (1983) Multiple criteria decision methods and applications. Springer, Berlin
13. Douligeris C, Pereira I.J (1994) A Telecommunications Quality Study Using the Analytic Hierarchy Process. IEEE J Sel Areas Commun vol 12
14. Tam M.C.Y, Tummala VMR (2001) An application of the AHP in vendor selection of a telecommunications system, vol 29. Omega, Elsevier, pp 171–182
15. Yeh C (2002) A problem-based selection of multi-attribute decision making methods. Int Trans Oper Res 9:169–181
16. Ariff H, Salit M.S, Ismail N, Nukman Y (2008) Use of analytical hierarchy process (AHP) for selecting the best design concept, vol 49
17. Simanaviciene R, Ustinovichius L (2010) Sensitivity analysis for multiple-criteria decision making methods: TOPSIS and SAW, Procedia—Social and Behavioral Sciences 2(6):7743–7744
18. Karami A (2011) Utilization and comparison of multi-attribute decision making techniques to rank Bayesian network options, Master Thesis, University of Skovde
19. Devi K, Yadav SP, Kumar S (2009) Extension of fuzzy TOPSIS method based on vague sets. Int Journal Comput Cognition 7(4)
20. Palanivel K, Sivakumar R (2010) Fuzzy multi-criteria decision-making approach for collaborative recommender systems. Int J Comp Theor Eng 2:1793–8201
21. Frair L, Matson JE (1998) Undergraduate curriculum evaluation with the analytic hierarchy process, Frontiers in Education Conference, IEEE
22. Hyun G-K, Seong P-H (1999) A methodology for evaluating alarm-processing systems using informational entropy-based measure and the analytic hierarchy process. IEEE Trans Nucl Sci 46:2269–2280
23. Cebeci U (2009) Fuzzy AHP-based decision support system for selecting ERP systems in textile industry by using balanced scorecard. Expert Syst Appl 36(5):8900–8909
24. Ozcan T, Celebi N (2011) Comparative analysis of multi-criteria decision making methodologies and implementation of a warehouse location selection problem. Expert Syst Appl Int J vol 38
25. Yang CL, Chuang SP, Huang RH, Tai CC (2008) Location selection based on AHP/ANP approach, Industrial Engineering and Engineering Management, IEEM, 2008
26. Rezaiana S, Joziba SA (2012) Health-safety and environmental risk assessment of refineries using of multi-criteria decision making method. APCBEE Procedia 2:235–238
27. Saeed Zaeri M, Sadeghi A, Naderi A (2011) Application of multi-criteria decision making technique to evaluation suppliers in supply chain management. Afr J Math Comput Sci Res 4 (3):100–106
28. Chang PL, Chen YC (1994) A fuzzy multi-criteria decision making method for technology transfer strategy selection in biotechnology. Fuzzy Sets Syst 63(2):131–139
29. Dagdeviren M, Yavuz S, Kilinc N (2009) Weapon selecting using the AHP and TOPSIS methods under fuzzy environment. An Expert Syst Appl 36(4):8143–8151

Book Recommender System Using Fuzzy Linguistic Quantifiers

Shahab Saquib Sohail, Jamshed Siddiqui and Rashid Ali

1 Introduction

One of the major contributions of the advancement in technologies is the ease of access it has facilitated to users for solving many daily life problems in fewer amounts of time and minimal efforts. The shopping has become easier through various online shopping portals, the people who are geographically away are easily connected, and it has become a matter of one click to explore the news from all over the world and get acquaintances with the views and ideas of the common people as well as the experts just by sitting at home and utilizing the modern technologies facilitated through TV, the Internet, etc. This lends the people to be aware of other's feeling and emotions, and being assisted with their ideas. Since it is a human tendency of relying upon the recommendation of their trusted persons, Recommender Systems (RS) have gained extraordinary popularities [1]. The state-of-the-art technologies help users to find the product and services as per their requirements [2]. Recommender system prior to recommendation identifies the behaviour of the users and makes use of their friends, neighbours and relatives recommendation or a man with similar choices to recommend them their desired items [3, 4].

It is assumed that recommender systems use customer's preferences and follow the philosophy that customers with similar taste have the similar choices while they shop. However, it may not be true for all the time. Let us consider selection of

S. S. Sohail (✉) · J. Siddiqui
Department of Computer Science, Faculty of Science,
Aligarh Muslim University, Aligarh 202002, Uttar Pradesh, India
e-mail: shahabsaquibsohail@gmail.com

J. Siddiqui
e-mail: jamshed_faiza@rediffmail.com

R. Ali
Department of Computer Engineering, Aligarh Muslim University, Aligarh 202002, India
e-mail: rashidaliamu@rediffmail.com

© Springer Nature Singapore Pte Ltd. 2017
R. Ali and M. M. S. Beg (eds.), *Applications of Soft Computing for the Web*,
https://doi.org/10.1007/978-981-10-7098-3_4

academic books which should not be dependent upon student's choice, rather it must be sincerely handled and the final decision by experts would be advisable. Therefore, the consideration of authorities' recommendations would be more effective. Hence, at the time of making recommendation for books, knowing the percept of experts at universities instead of common people seems more appealing which can help in avoiding these issues.

In this chapter, we intended to recommend top books for university's students in Indian perspectives. That is why we have chosen top-ranked universities from India. The selection of top-ranked universities is based upon the QS world university ranking. QS ranking is one of the leading rankers of the academic institution. Once the top institutions are explored, their syllabus for the particular subject is searched which served as a base for the recommendation of books. Positional Aggregation Scoring (PAS) technique [5] is highly advisable for aggregating a final result from these types of data. Finding best amongst different books by ranked university is supposed to be more helpful, as it shortens the data overload, reduces the complexities and increases the authenticity of the products. PAS is simple aggregation method; we use Ordered Weighted Averaging (OWA) operator to transform this simple aggregation method into weighted aggregation.

OWA is a fuzzy-based averaging operator, which in combination with linguistic quantifier gives a variation of option for decision-making problems. By using different linguistic quantifiers, we get different weights which can be assigned to universities that give the idea about the involvement of the university in the recommendation process, e.g. 'most' is a linguistic quantifier; by using it, those books are preferred for which 'most' of the universities have high ranking. In the same way, scores for books are obtained using different quantifiers. The top-ranked books are recommended as a final recommendation for the users. These ranked books are compared with the ranking taken from the experts. P@10, Mean Average Precision (MAP) and FPR@10 are used as evaluation metrics to check the performance of the proposed scheme. Further, upon the basis of the expert's ranking, the comparison of the proposed approach is made with books recommended by amazon.com for the same courses. It is assessed that what is the suggestion from amazon.com and proposed scheme with comparison to ranking of the experts, assuming that ranking by the experts is standard. The results show that the proposed scheme performs slightly better than Amazon recommendations.

Rest of the paper is organized as follows. In Sect. 2, we have illustrated examples to give a background of the proposed approach. Section 3 explains the book recommendation approach with suitable examples. The elaborative experimental results are discussed in Sect. 4. Section 5 concludes with some future directions.

2 Background

This section deals with basic concept involved in aggregation problems which enable users to understand these problems better. Further, product recommendations are also discussed with an inclusive discussion on book recommendation.

The authors have [6–8] utilized user feedback for several electronics items in recommendation process. Liu et al. [9] combined group decision-making and data mining for product recommendation.

There are several product's recommendation works that have been reported in literature [10–12]; however, the book recommendation is still to be explored adequately in the literature. A book recommendation technique known as LIBRA (Learning Intelligent Book recommendation Agent) has been proposed by Mooney [13]. This technique is based on the content filtering which also employs machine learning methods for recommending books. Association rule has also been used in book recommendation; one of these works has been reported by Jomsri [14]. The author has recommended book for library which takes consideration of user profiles, too. However, the technique is designed to benefit the users located in same geographical area, i.e. residents of same hostel of a university or same locality. Also, the author has not performed the experiment for geographically distant institutes; rather it works for the students of same institute only.

The authors [15] have also used association rule mining with the combination of collaborative filtering and classification. By using methods for classification, the feature-based classification of books is done. The associated rule mining and collaborative filtering are used to recommend those books which have been rated high by the users and that match the requirements of the users too. Digital signage system is used to recommend the books by the authors in [16]. They identify age and sex of the users prior to making any recommendation. However, the scope of the recommendation is very precise and limited. The adopted method cannot be generalized and a bigger community cannot get benefit from it. Instead, the method is designed for the users whose age span between 19 and 21 years; moreover, they are also located in close region. In [8, 17, 18], book recommendations based on soft computing techniques have been reported.

The authors in [19] have categorized the feature for the online opinion by the users to recommend books. The different computer science books are assigned weights and accordingly scored. The authors have used Google search to explore the book on the topic of interest. The links which appear high in the SERP are stored. The assessment of the stored reviews is performed to extract the features. The books are ranked on the basis of the features which are extracted from the reviews of the users. The use of OWA for the book recommendation has been performed in [20]. The authors have implied soft computing technique for the recommendation of the books for university graduates; however, the authors have used only one subject of the book to recommend and they have not shown any comparison of their approach with any other methods. Kim et al. [21] have proposed a book recommender system for the validation of their method which was designed for an online community. They tried to satisfy the minor members of group which are left unsatisfied although the majority may have satisfaction due to the differences in preferences. In this chapter, we have tried to overcome the shortcomings of the above approaches. The main contributions in the chapter can be summarized as follows:

- We incorporated aggregation of experts' decision to recommend books. This enables to visualize the book recommendation problems as a decision-making problem.
- Positional Aggregation based Scoring (PAS) Technique with the combination of OWA is implemented for recommendation process.
- The concept of fuzzy by the use of linguistic quantifier with the support of Ordered Weighted Aggregation (OWA) is employed for book recommendation. As far as our search is concerned, we did not find any book recommendation approach employing these techniques for the said techniques.
- Comparison of the proposed scheme is shown with the amazon.com.

3 Book Recommendation Approach

The basis of our approach is the fact that the best-ranked university will have most appropriate recommendation of books to students. For this, the top-ranked university in India from best ranking site is taken. Further, soft computing techniques have been used to aggregate the ranking of the universities with the consideration of their ranks and the ranking of the books by corresponding universities. Again, the ranking of the books of concerned topic is taken from experts of the subject from reputed universities. These rankings are used to evaluate the adopted approach. Architecture for the proposed approach and its evaluation is shown in Fig. 1.

3.1 Selection of Top Universities and Recommended Books

The selection of top universities can fulfil the purpose of recommendation of appropriate books for graduate students as far as methodology is concerned. The books which are recommended by top institutions should be more reliable for students rather than the books recommended by other corporate recommender sites. The ranking of QS World University Rankings in collaboration with Elsevier is considered for the procedure that ranges about 40 subjects all around the globe.

QS World University Rankings for computer science and information systems has only seven institutions from India. We have selected these institutions for the inclusion of books. The list of top Indian institution for 2015 is listed below in Table 1.

We have obtained 158 different books of 10 different courses. Due to the limitations of pages, only 3 different courses containing 47 distinct books are considered. Out of these combinations, we intended to present the top 10 ordered books of each course before users.

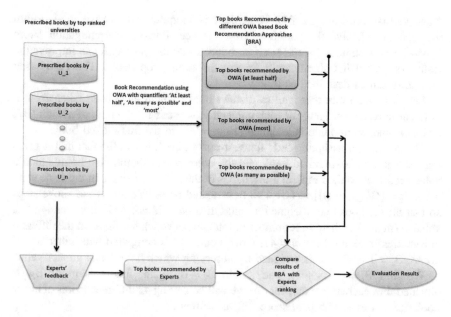

Fig. 1 Architecture of the book recommendation using OWA

Table 1 Top 7 Indian universities in QS ranking [32]

Rank position	1	2	3	4	5	6	7
University name	IIT, Bombay	IIT, Delhi	IIT, Kanpur	IIT, Madras	IISc, Bangalore	IIT, Kharagpur	IIT, Roorkee

3.2 Positional Aggregation Based Scoring Technique (PAS)

We are aimed at suggesting the books for the syllabus of top-ranked universities. As discussed in the above section, the course curriculums of the different universities notably differ from each other. Thus, it gives us a partial list of the books recommended in the syllabus by different universities. The ranking of final aggregated books has been obtained by using aggregation algorithm. This list may be partial as well full. The Borda's method [22] is one of the most used methods for the purpose. There are other various prominent techniques like Markov chain based methods [23] and soft computing based methods. These techniques have been designed to work well with full list [22, 24]. Thus, there is a requirement to have an algorithm that can be effective for partial list, too. Hence, Positional Aggregation Score (PAS)-based technique [5] has been introduced into the picture. It can improve the result for partial list significantly for recommending top books. The PAS-based technique can be described as follows:

Let us assume that there are m distinct books which have been taken from n different universities. First, for each university, we seek for the rank of a book 'B_i', and then maximum value is assigned ($V_{max} = -1$) to the first-ranked book.

The concept of assigning '-1' for best-ranked book lies in the fact that it gives highest value to best-ranked book and the other books should also be assigned a value according to their rank in the syllabus of the top universities. In the same way, we assign $\{(V_{max}) - (i)\}$ to $(i + 1)$th best-ranked book. We repeat the above steps so that all the books are assigned a value. If a book is not ranked, we assign it a value '$-(m + 1)$', where m is number of total books which has fallen in the different universities' ranking, here $m = 41$. Now, book 'B_i' is compared with other '$m-1$' books. $B_i = 1$ or $B_i = 0$ is assigned by observing which book has a greater value.

We find $B_i == -(m + 1)$, and assign $B_i = 0$, again; finally, sum of B_is is obtained. For each university, for this we will be getting 42 different scores of every book, termed as S. The final score 'FS' is written as

$$FS = (S/(m+1)).$$

Tables 2, 3 and 4 illustrate the above example for book 'compiler design' having book code 'CD'.

Table 2 Ranking of books 'compiler design' by top 7 universities

Book code	U_1	U_2	U_3	U_4	U_5	U_6	U_7
CD.1.	1	3	1	1	1	0	1
CD.2.	2	1	2	0	0	0	0
CD.3.	3	0	0	0	0	0	0
CD.4.	0	2	7	0	2	0	0
CD.5.	0	0	3	0	0	0	3
CD.6.	0	0	4	0	0	0	4
CD.7.	0	0	5	0	0	0	0
CD.8.	0	0	6	0	0	0	0
CD.9.	0	0	0	0	0	0	2
CD.10.	0	0	0	0	0	0	5

Table 3 Intermediate steps of PAS calculation

Book Code	U_1	U_2	U_3	U_4	U_5	U_6	U_7
CD.1.	−1	−3	−1	−1	−1	0	−1
CD.2.	−2	−1	−2	0	0	0	0
CD.3.	−3	0	0	0	0	0	0
CD.4.	0	−2	−7	0	−2	0	0
CD.5.	0	0	−3	0	0	0	−3
CD.6.	0	0	−4	0	0	0	−4
CD.7.	0	0	−5	0	0	0	0
CD.8.	0	0	−6	0	0	0	0
CD.9.	0	0	0	0	0	0	−2
CD.10.	0	0	0	0	0	0	−5

Table 4 Final PAS score for book 'Compiler Design'

Book Code	U_1	U_2	U_3	U_4	U_5	U_6	U_7	PAS
CD.1.	1.00	0.78	1.00	1.00	1.00	0.00	1.00	0.83
CD.2.	0.89	1.00	0.89	0.00	0.00	0.00	0.00	0.40
CD.3.	0.78	0.00	0.00	0.00	0.00	0.00	0.00	0.11
CD.4.	0.00	0.89	0.33	0.00	0.89	0.00	0.00	0.30
CD.5.	0.00	0.00	0.78	0.00	0.00	0.00	0.78	0.22
CD.6.	0.00	0.00	0.67	0.00	0.00	0.00	0.67	0.19
CD.7.	0.00	0.00	0.56	0.00	0.00	0.00	0.00	0.08
CD.8.	0.00	0.00	0.44	0.00	0.00	0.00	0.00	0.06
CD.9.	0.00	0.00	0.00	0.00	0.00	0.00	0.89	0.13
CD.10.	0.00	0.00	0.00	0.00	0.00	0.00	0.56	0.08

3.3 Ordered Weighted Aggregation (OWA)

Ordered Weighted Aggregation (OWA) is introduced to handle uncertainty in decision-making problem where number of criteria is well defined. It was proposed by Yager in 1988 [25].

It has been used in literature for diverse field of application [26] which includes novel fuzzy queries for web searching [27], GIS applications [28, 29], talent enhancement of sportsperson [30], etc.

Ordered Weighted Aggregation (OWA) is mathematically expressed as

$$OWA(y_1, y_2, \ldots, y_n) = \sum_{k=1}^{n} w_k z_k, \tag{1}$$

where Z_k refers that re-ordering of $y_1, y_2, \ldots, y_n$ in descending order will result a sequence $z_1, z_2, \ldots, z_n$ such that $z_n \leq z_{n-1} \ldots \leq z_2 \leq z_1$. The weights '$W_k$' can be obtained with the help of the following equation [21, 29, 31]:

$$W_k = \{Q\,(k/m) - Q((k-1)/m)\}, \tag{2}$$

where $k = 1, 2, \ldots m$.

Function $Q(r)$ for relative quantifier can be calculated as

$$Q(r) = \begin{cases} 0 & \text{if} \quad r < a \\ \frac{(r-a)}{(b-a)} & \text{if} \quad a \le r \le b \\ 1 & \text{if} \quad r > b \end{cases} \tag{3}$$

$Q(0) = 0$, $\exists r \ \varepsilon \ [0, 1]$ such that $Q(r) = 1$, such that a, b and $r \ \varepsilon \ [0,1]$ which are parametric values.

The distinct weights will be obtained by using different linguistic quantifiers, e.g. 'most', 'at least half', etc.; considering these linguistic quantifiers helps in assessing the behaviour of university towards particular book, which is a linguistic quantifier, by using it those books are preferred in which 'most' of the university consider books in their syllabus. Thus, all the books are sorted and ranked based upon the values they have obtained.

Example 3.2: For number of criteria $(m) = 5$ and parametric values as a = 0 and b = 0.5, we will have corresponding weights for OWA values as

Values of weights	Quantifiers values: $a = 0, b = 0.5$	Quantifiers values: $a = 0.3, b = 0.8$
$w\,(1)$	0.4	0.0
$w\,(2)$	0.4	0.2
$w\,(3)$	0.2	0.4
$w\,(4)$	0.0	0.4
$w\,(5)$	0.0	0.0

$$\text{OWA} = \sum_{k=1}^{n} w_k z_k$$

$$\begin{aligned} \text{OWA (as many as possible)} &= 1 * 0 + 0.75 * 0 + 0.6 * 0 + 0.5 * 0.14285 + 0.4 * 0.28571 \\ &\quad + 0 * 0.28571 + 0 * 0.28571 \\ &= 0.185709 \end{aligned}$$

Similarly, for seven criteria with parametric values (0, 0.5) and (0.3, 0.8), we will be getting weights as $W(1) = 0.28571$, $W(2) = 0.28571$, $W(3) = 0.28571$, $W(4) = 0.14285$, $W(5) = 0.0$, $W(6) = 0.0$ and $W(7) = 0.0$, and $W(1) = 0.0$, $W(2) = 0.0$, $W(3) = 0.25714$, $W(4) = 0.28571$, $W(5) = 0.28571$, $W(6) = 0.17142$ and $W(7) = 0.0$, respectively.

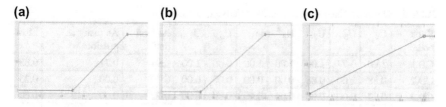

Fig. 2 **a** As many as possible, **b** most, **c** at least half

A graphical representation of these fuzzy linguistic quantifiers is shown in Fig. 2a–c, respectively, for 'as many as possible', 'most' and 'at least half', respectively.

The basis of the whole recommendation process adopted is the top universities in the QS ranking and their recommended books for enrolled student at respective campuses. In PAS technique, we have not assigned any weights to the universities. Hence, PAS can be perceived as un-weighted aggregation of scores assigned to books. Each course which consists of several books is assigned a score using PAS. By using OWA with different linguistic quantifiers, we can assign weights to the university. The weighted aggregation of books may lead to a better list of top books.

The weights are assigned to universities using linguistic quantifier 'most', 'at least half' and 'as many as possible', and applying OWA is listed in Table 5. These weights including PAS score of books of the different courses for respective universities are applied in equation 1 for obtaining OWA scores of books which in turn gives top books, after sorting as shown in Table 6. These books are ranked and recommended.

4 Experiments and Results

In the previous section, it is shown that how the books are scored and ranked. Now, we take explicit feedback from experts about the rankings of books. Explicit feedback means the users are given the total books of the three different subjects

Table 5 Weights assigned to universities using relative quantifiers

Ranked universities		U_1	U_2	U_3	U_4	U_5	U_6	U_7
Weights of relative qualifier	Most	0	0	0.28571	0.28571	0.0711	0.1714	0
	At least half	0.28571	0.28571	0.28571	0.14285	0	0	0
	As many as possible	0	0	0	0.1428	0.28571	0.28571	0.28571

Table 6 Scores obtained by books using relative quantifiers

Book code	U_1	U_2	U_3	U_4	U_5	U_6	U_7	At least half	As many as possible	Most
CD.1.	1.00	0.77	1.00	1.00	1.00	0.00	1.00	0.94	0.71	0.83
CD.2.	0.88	1.00	0.88	0.00	0.00	0.00	0.00	0.79	0.00	0.23
CD.3.	0.77	0.00	0.00	0.00	0.00	0.00	0.00	0.22	0.00	0.00
CD.4.	0.00	0.88	0.33	0.00	0.89	0.00	0.00	0.35	0.25	0.34
CD.5.	0.00	0.00	0.78	0.00	0.00	0.00	0.78	0.22	0.22	0.20
CD.6.	0.00	0.00	0.67	0.00	0.00	0.00	0.67	0.19	0.19	0.17
CD.7.	0.00	0.00	0.56	0.00	0.00	0.00	0.00	0.16	0.00	0.14
CD.8.	0.00	0.00	0.44	0.00	0.00	0.00	0.00	0.13	0.00	0.11
CD.9.	0.00	0.00	0.00	0.00	0.00	0.00	0.89	0.00	0.25	0.00
CD.10.	0.00	0.00	0.00	0.00	0.00	0.00	0.56	0.00	0.16	0.00

and asked to give their top 10 recommendations. These users are expert academia persons and corporate giants who have experiences in the concerned fields. Three different experts are approached for each course. We treat these rankings as the basis for further evaluation. Also, the top 10 rankings of these books in amazon. com are compared with the base ranking. The results are evaluated on the basis of P@10, FPR@10 and MAP. The definition and results of the evaluation metric are discussed below.

4.1 P@10

We define the precision at top 10 positions as P@10 and it is given as

P@10 = (Number of books recommended in top 10 positions that are also preferred by experts)/10;

The value of P@10 for different linguistic quantifiers is shown in Fig. 3. It is evident from the figure that while using at least half as quantifier we are obtaining three times more precise results than amazon.com.

4.2 Mean Average Precision (MAP)

The mean average precision (MAP) for an 'n' user is the average of the average precision of each user, which is given as

$$\text{MAP} = \frac{1}{n}\sum_{i=1}^{n}\{P(Ui)\}, \tag{4}$$

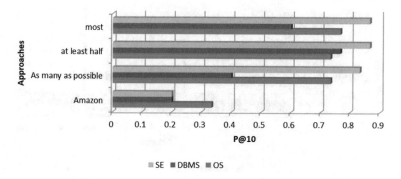

Fig. 3 Comparison of P@10 for different approaches

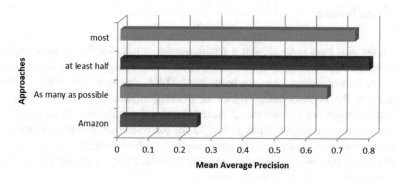

Fig. 4 Comparison of MAP for different approaches

where $P(U_i)$ is the precision of ith user. The MAP for the proposed scheme with all three quantifiers and their comparison is shown in Fig. 4.

4.3 FPR@10

We define the false positive rate at top 10 positions as FPR@10 and it is given as

FPR@5 = (Number of books recommended in top 10 position but not preferred by experts)/10.

The false positive rate is found to be minimum for the quantifier 'at least half', whereas amazon.com has higher false positive, as shown in Fig. 5.

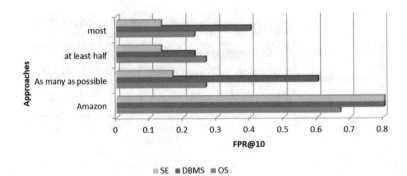

Fig. 5 Comparison of FPR@10 for different approaches

5   Conclusion

The basic idea here is to present a recommender system for books by applying a fuzzy-based aggregation operator, 'OWA', and employing aggregation of universities recommendations. As we have used OWA operator with three different quantifiers, the corresponding values of evaluation metric P@10 and MAP for ranking of books are shown, and it turned out that the quantifier 'at least half' has greater value of precision. Also, FPR@10 for the aforesaid quantifier is least.

The comparison of the results with Amazon ranking indicates that the recommendation of proposed approach is preferred by the expert users three times more than the ranking done by amazon.com.

Further, these techniques can be integrated to implement it over a larger data set and to a variety of books for respective users. Also, we may expand number of users to include them from various locations of the globe to verify the method more accurately.

References

1. Bobadilla J, Ortega F, Hernando A, Gutiérrez A (2013) Recommender systems survey. Knowl Based Syst 46:109–132
2. Adomavicius G, Tuzhilin A (2005) Toward the next generation of recommender systems: a survey of the state-of-the-art and possible extensions. IEEE Trans Knowl Data Eng 17 (6):734–749
3. Burke R (2007) Hybrid web recommender systems. Springer, Berlin, pp 377–408
4. Burke R, Felfernig A, Göker MH (1997) Recommender systems: an overview, AI magazines, 32(3). pp. 13–18
5. Ali R (2013) Pro-Mining: Product recommendation using web based opinion mining. Int J Comput Eng Tech 4(6):299–313
6. Sohail SS, Siddiqui J, Ali R (2014) User feedback scoring and evaluation of a product recommendation system. In: In contemporary computing (IC3), 2014 seventh international conference on IEEE, pp 525–530

7. Sohail SS, Siddiqui J, Ali R (2015) UMW: a model for enhancement in wearable technology based on opinion mining technique. In: 12th international conference on learning and technology, Jeddah, KSA, © IEEE, April 2015, pp 9–13
8. Sohail SS, Siddiqui J, Ali R (2015) User Feedback Based Evaluation of a Product Recommendation System Using Rank Aggregation Method. Springer International Publishing, In Advances in Intelligent Informatics, pp 349–358
9. Liu DR, Shih YY (2005) Integrating AHP and data mining for product recommendation based on customer lifetime value. Inf Manag 42:387–400
10. Weng SS, Liu MJ (2004) Feature-based recommendation for one-to-one marketing. Expert Syst Appl 26(4):493–508
11. Cheung KW, Kwok JT, Law MH, Tsui KC (2003) Mining customer product ratings for personalized marketing. Decis Support Syst 35(2):231–243
12. Goldberg D, Nichols D, Oki BM, Terry D (1992) Using Collaborative Filtering to Weave an Information TAPESTRY. Commun ACM 35(12):61–70
13. Mooney RJ, Roy L (2000) Content-based book recommending using learning for text categorization. In: Proceedings fifth ACM conference on digital libraries, San Antonio, USA, 2–7 June 2000, pp 195–204
14. Jomsri P (2014) Book recommendation system for digital library based on user profiles by using association rule. In: Proceedings 2014 fourth international conference on innovative computing technology (INTECH), 13–15 August 2014, pp 130–134
15. Tewari AS, Priyanka K (2014) Book recommendation system based on collaborative filtering and association rule mining for college students. In: Proceedings 2014 international conference on contemporary computing and informatics (IC3I), 27–29 Nov 2014, pp 135–138
16. Mikawa M, Izumi S, Tanaka K (2011) Book recommendation signage system using silhouette-based gait classification. In: Proceedings 10th international conference on machine learning and applications, pp 416–419
17. Sohail SS, Siddiqui J, Ali R (2014) Ordered ranked weighted aggregation based book recommendation technique : a link mining approach. In: 14th international conference on hybrid intelligent systems (HIS), IEEE, pp 309–314
18. Sohail SS, Siddiqui J, Ali R (2016) Book recommender system using fuzzy linguistic quantifier and opinion mining. In: The international symposium on intelligent systems technologies and applications, Springer International Publishing, pp 573–583
19. Sohail SS, Siddiqui J, Ali R (2013) Book recommendation system using opinion mining technique. In: Advances in Computing, communications and informatics (ICACCI), 2013 international conference on IEEE, 2013, pp 1609–1614
20. Sohail SS, Siddiqui J, Ali R (2015) OWA based Book Recommendation Technique. Procedia Comput Sci 62:126–133
21. Kim JK, Kim HK, Oh HY, Ryu YU (2010) A group recommendation system for online communities. Int J Inf Manag 30:212–219
22. Borda JC (1781) Memoire sur les election au scrutiny. Histoire de l'Academie Royale des Sciences
23. Dwork C, Kumar R, Naor M, Sivakumar D Rank aggregation methods for the web. In Proceedings of the tenth international conference on world wide web, Hong Kong, 1–5 May, pp 613–622
24. Beg MMS, Ahmad N (2003) Soft computing techniques for rank aggregation on the World Wide Web. World Wide Web Int J 6(1):5–22
25. Yager R (1988) On ordered weighted averaging aggregation operators in multicriteria decision Making. IEEE Trans Syst Man Cybern 18(1):183–190
26. Beliakov G, Pradera A, Calvo T (2007) Aggregation functions: a guide for practitioners. Springer, Heidelberg
27. Beg MMS (2005) User feedback based enhancement in web search quality. Inf Sci 170(2–4):153–172

28. Malczewski J (2006) Ordered weighted averaging with fuzzy quantifiers: GIS-based multicriteria evaluation for land-use suitability analysis. Int J Appl Earth Obs Geoinf 8 (4):270–277
29. Makropoulos CK, Butler D (2006) Spatial ordered weighted averaging: incorporating spatially variable attitude towards risk in spatial multi-criteria decision-making. Environ Model Softw 21(1):69–84
30. Ahamad G, Naqvi SK, Beg MM (2015) An OWA-based model for talent enhancement in cricket. Int J Intell Syst 31(8):763–785
31. Sohail SS, Siddiqui J, Ali R (2017) A novel approach for book recommendation using fuzzy based aggregation. Indian J Sci Tech 8:1–30
32. http://www.topuniversities.com/university-rankings/university-subject-rankings/2015/computer-science-information-systems#sorting=rank+region=+country=96+faculty=+stars=false+search=
33. Sohail SS, Siddiqui J, Ali R An OWA based ranking approach for university books recommendation. Int J Intel Syst https://doi.org/10.1002/int.21937 (in press)

Use of Soft Computing Techniques for Recommender Systems: An Overview

Mohammed Wasid and Rashid Ali

1 Introduction

Today, the Internet and users are growing rapidly and generate huge amount of data over the Internet, which has caused both seller and customers to face information overload problem. As there are a large number of products available over the Internet and it is really difficult for a user to access all products at a time. The seller wants to expand their revenue by selling more products and the customer wants to purchase desired items in lesser time. Therefore, researchers from both the industry and academia are trying to develop new intelligent techniques to overcome the information overload problem by providing more relevant products to the customers from large amount of product list. One possible solution for such intelligent technique is the use of Recommender Systems (RSs) [1]. RSs are the rapidly emerging information filtering software tool, which identifies interesting products and items by considering the customer's past or previous preferences and increase customer satisfaction and faith in e-commerce purchase. These systems reduce the e-commerce time complexity problem and make the product selection job easier by retrieving only useful set of items from a large number of item set. Before making recommendations, the RSs first analyze the user and product information. This information can be obtained via explicit or implicit mechanism. Data which is collected by asking users for information comes under the explicit information category, while data collected by observing the user's behavior comes under the implicit information category [2].

M. Wasid (✉) · R. Ali
Department of Computer Engineering, Aligarh Muslim University,
Aligarh 202002, Uttar Pradesh, India
e-mail: erwasid@gmail.com

R. Ali
e-mail: rashidaliamu@rediffmail.com

© Springer Nature Singapore Pte Ltd. 2017
R. Ali and M. M. S. Beg (eds.), *Applications of Soft Computing for the Web*,
https://doi.org/10.1007/978-981-10-7098-3_5

After obtaining this input information, the RSs use any of its filtering technique to generate recommendations for users. Generally, RSs are classified into three different recommendations techniques for generating recommendations [3] namely; content-based recommenders (CBR), collaborative filtering recommenders (CFR), and hybrid filtering recommenders (HFR). The CBR use product's information (content or features) to generate recommendations while CFR uses customer's similar users (now onwards we call them as "neighbors") for making recommendations. HFR generate recommendations by combining CBR and CFR techniques together. Each recommendation technique has its own advantages and limitations; for example, CBR faces overspecialization problem, while CFR faces data sparsity, scalability, and cold-start problems. In recent years, researchers are investigating the use of different Soft Computing (SC) techniques in the area of recommendation systems for providing more effective suggestions to users by alleviating these problems. Soft Computing [4] became a formal area of study in computer science in the early 1990s and used to handle different challenges offered by data mining [2] and other major areas. SC techniques have also been widely used in the development and implementation of different recommendations methods to overcome some of the limitations of existing recommendations techniques [5]. These techniques have shown significant potentials to make RSs more robust, effective, and accurate. Basically, SC is the consortium of methodologies that are aimed to provide, in one form or another, efficient solutions to the real-life problems which are not modeled or too difficult to model, mathematically. The primary aim of these techniques is to exploit the understanding of imprecision, approximate reasoning, uncertainty and partial truth to undertake robustness, tractability and low solution cost, and to mimic human decision-making strategy.

Generally, SC techniques include fuzzy sets (FS), neural networks (NN), evolutionary computing (EC), and swarm intelligence (SI) as shown in Fig. 1. All these techniques are briefly discussed in the following part.

Lotfi Zadeh in 1965 introduced the concept of fuzzy sets, which does not deal with the crisp set and well suited for handling the incomplete or imprecise information [6]. Fuzzy logic mimics the human decision-making pattern which is generally used in controller systems and allows products to have partial degree of membership (i.e., result in the interval [0, 1]) instead of hard membership of the products to the crisp set (i.e., 0 if product does not belong to the set, and 1 for otherwise). Fuzzy RSs can be developed by proposing an appropriate relationship function or method, to compute the similarity between users and preparing a fuzzy user model by handling the uncertain user-item features in the system.

Neurons are very simple processing elements in the brain, connected with each other in a massively parallel fashion [4]. NN is developed to mimic the human brain intelligence and discriminating power. A simple NN model contains two layers: input layer and output layer; while Backpropagation Network (BPN) contains an additional intermediated hidden layer in between input and output layers. NN is trained by adjusting the weights of the connections between elements with the help of training data examples [7]. The training of NN is done in such a way that a particular input information leads to a specific target output. Back Propagation

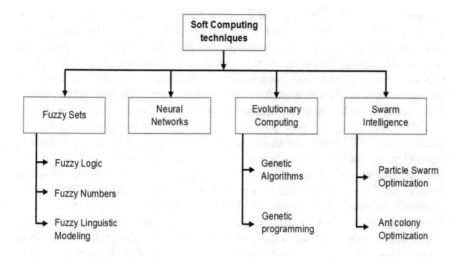

Fig. 1 Soft computing techniques

(BP) algorithm is a training algorithm designed to minimize the deviation between the actual and desired objective function values.

Several recommendation approaches have been proposed for the use of Evolutionary Computing (EC) to learn user profile features for generating more accurate results to the users [5, 8]. Genetic Algorithm (GA) is a stochastic search technique which can provide solutions for diverse optimization problems with efficiency on complicated and large search spaces [3]. GA works on Darwin's principle of "survival of the fittest" and gives a method to model biological evolution. It starts with a population of individuals called chromosomes, where each chromosome has a number of genes and used to decide the possibility of reproduction for the next generation on the basis of its fitness value. The chromosome which has best (fittest) value will be used to form new population iteratively [9]. Three genetic operators are responsible for constructing the new generations namely, selection, crossover, and mutation. The initial set of chromosomes is selected for the new generation using selection operator based on their fitness values. Whereas, the crossover is used to generate two new offsprings by swapping a portion of two selected parent chromosomes. Mutation operator produces offspring by randomly altering the status of a gene bit. At the end, the algorithm terminates when a predefined number of generations reached or fitness threshold is met.

Swarm intelligence (SI) [2] is used to solve complex optimization problems through the principles of self-organization, decentralization, and cooperation through communication. SI concept was inspired from the swarm of social organisms like birds, fish, and animals. There are many different types of SI techniques but Particle Swarm Optimization (PSO) and Ant Colony Optimization (ACO) are the two main techniques widely used for solving the recommendation problems in the past. Kennedy and Eberhart introduced the biologically inspired heuristic search algorithm called PSO algorithm for solving the optimization problems [10]. PSO mimics the social behavior of a flock of birds randomly flying in a multidimensional search space in search of food. Therefore, in PSO each bird is represented as a particle and has initialized with a random position and velocity through which they flew through the problem search space. The location or position and velocity of each particle get updated based on its previous best position (pbest), and the previous global best position of a particle in the population (gbest). Consequently, PSO keeps accelerating each particle by updating their position and velocity using the "pbest" and "gbest" until a certain threshold is met or a predefined number of generation is reached [11]. Similar to PSO, ACO comes under the swarm intelligence class which was proposed by Dorigo [12] to solve the hard combinatorial optimizations problems. The idea of ant colony optimization is as its name suggests, inspired by the foraging behavior of the living ants in the environment. The main approach of ACO is the chemical substance, which is produced by ants during the search of food, known as pheromone. The pheromone updating strategy is used for solving the optimization problems. The path which has maximum pheromone is the one that has been visited recently by the ants and hence the most suitable for following ants to visit. Ants will follow the most optimized path based on the strength of the pheromone and pheromone of other path is evaporated over time such that as time proceeds.

In this paper, we surveyed the soft computing-based recommender systems for each technique separately. Although, there are many existing RSs survey articles available, they mainly focus on recommendation theories and approaches and best to our knowledge there is no such survey article available which provides a particular review for RSs comprising soft computing techniques for making recommendations. We have perceptively and comprehensively summarized recent research on recommender systems from the soft computing techniques point of view, which leads to a soft computing-based framework for recommender system development. This survey covers recent recommendations and approaches separately for each soft computing technique with several possible research directions for each technique which will directly motivate and support researchers and practitioners to further explore the area.

The remainder of this paper is organized as follows. Section 2 discusses the basic concept of recommender systems and some of the recommendation techniques are reviewed. In Sect. 3, we presented some recent soft computing techniques-based recommender system and also highlights the associated challenges and future research scope discussed. A soft computing-based recommender systems framework is discussed and presented in Sect. 4. Section 5 concludes the survey.

2   Recommendation Techniques

In order to clearly understand and analyze the key mechanism of recommender systems, in this section, we first discuss recommender systems and then we will review different recommendation techniques, like traditional techniques such as content-based recommender, collaborative filtering recommender, and hybrid filtering recommender [13].

Recommender Systems have become an increasingly important research area since the mid-1990s when researchers started focusing on developing RSs techniques and methodologies for various domains like movie, music, books, etc. RSs are a web personalization information filtering tool that is used to assist users by providing relevant product recommendations at the right time [1]. Normally, RSs generate recommendations using a set of input data. Input data can be collected through explicit and implicit data method. Normally, explicit ratings are binary in nature (i.e., like or dislike) or have some specified numeric scale to represent the likeness for a product (i.e., 1—bad to 5—excellent), where 5 is the maximum rating supported by the system [14].

I. Content-based recommender (CBR)

CBR generates recommendations to a user based on the contents of the item preferred by the user in the past. The recommendations mainly depend on the features of the item and past preferences of the user, for e.g., purchase history, search query, and ratings [15]. Traditional CBR initially builds user profile using user ratings and features of items. Many learning techniques like Bayesian networks have been used for building a user profile. This user profile when compared with other unexplored item's features and best matching items are recommended to the user [16]. For instance, LIBRA [17] generate recommendations for books using book features taken from Amazon web pages. It first builds a user profile and generates recommendations using book ratings plus description extracted from Amazon web pages. Different profile-item matching techniques are used by CBR method to compare each item's attributes with the user profile and decide whether this item is relevant to the target user or not based on the degree of similarity between them. The similarity can be obtained using cosine similarity, keyword matching, and typical classifications methods. There are two different techniques used for making recommendations. First, heuristic technique generates recommendations using traditional information retrieval methods, for instance, Pearson similarity method [18]. The second technique produces recommendations using machine learning and statistical learning methods [15].

II. Collaborative filtering recommender (CFR)

Collaborative filtering recommenders are depended upon the preferences of a set of similar users to the target user. These similar users share common tastes on a set of items to the target users in the past. For instance, for watching a movie, one may ask her friends' opinion about the movie. Usually, CFR is divided into three main categories, namely, memory-based, model-based, and hybrid technique [1, 13].

Memory-based algorithm also known as neighborhood-based algorithms are the main algorithms developed for CFR. The key mechanism for these algorithms is based on the user ratings given to the items. These ratings are stored in the form of a user-item matrix where rows indicate the users and columns indicate the items rated by the users. The memory-based algorithms are further divided into the user-based and item-based algorithms. The user-based algorithms work on the basis of similarity among the users where a group of most similar users is formed, for making predictions. Similarity among the users is calculated using any of the similarity measures, for e.g., Pearson correlation measures the degree of correlation between two profiles. Whereas item-based algorithms works on the basis of similarity among the items and most similar items form a group. After creating the group of similar items, ratings of the neighbor item is used to predict the rating of the unrated item.

The next category of the recommender system is the model-based algorithm, which generates recommendations in two phases. First, a model is created by using the rating matrix and then this model is used to generate recommendations to users. The model can be prepared with the help of any data mining or machine learning algorithms. For example, regression model, classification, clustering algorithms can be used for model building. The third category of CFR is hybrid algorithms which combine both memory-based and model-based algorithms for generating more effective recommendations to users. Generally, hybrid algorithms are used to overcome the limitations of both the algorithms by combining them into a single compact model.

III. Hybrid filtering recommender (HFR)

CBR and CFR have their specific strength and limitations, therefore, HFR combines both these techniques to achieve better recommendation accuracy. Researchers have shown that combinations of these two techniques into one would alleviate many usual problems faced by traditional recommender systems such as cold-start, data sparsity, and scalability problems [19]. There are mainly three ways to implement HFR approach; (1) apply simple CBR technique and then implement CFR technique, (2) apply CFR technique and then implement CBR technique, (3) implement both techniques together through some modeling. The system proposed by Robin Burke [19] is an example of the hybrid system, which comprises both content-based and collaborative filtering techniques.

3 Recommender Systems Based on Soft Computing Techniques

Soft computing techniques are effectively employed in a variety of domains and showing fairly immense potential for the improvement of RSs [4, 5, 20]. There are very few real datasets, which are openly available for RSs research. Moreover, most of them are sparse, incomplete, and noisy in nature. For example, the sparsity level of widely popular 1 M MovieLens dataset is 0.9369 and sparsity level of Netflix dataset is 0.9882 [5, 21]. Most of the problems in recommender systems occur due to lack of user-item rating and impreciseness of the data [11, 14]. Therefore, researchers are trying to incorporate different approaches for handling these problems in RSs. Soft computing techniques seem to be a very useful approach for representing and managing impreciseness of data [14]. In the below subsections, we will discuss some recent developments in RSs based on each soft computing techniques, separately, used for improving the prediction accuracy of recommendations.

(a) Recommender systems based on Fuzzy Logic

Fuzzy logic has grown considerably in the field of recommender systems and in other wide variety of RSs domains such as, in product recommendations [18, 22], semantic music recommendation system [23], etc. Fuzzy set offers a rich spectrum of methods by dealing with the imprecise, incomplete, and uncertain information. For example, item features in content-based recommenders are represented using binary representations but binary representation of item's features fails to represent real life of feature presence. Therefore, fuzzy concept is used here to denote real-life scenario by representing item features using fuzzy logic. In CFR, the recommendation quality depends upon the similarity function used to generate neighborhood set for the target user. A good (bad) similarity method will generate good (bad) neighborhood set which results in good (bad) recommendation accuracy. A bad recommendation will lead to reduce the trust of users on the RSs therefore, it is a major concern to choose these similarity methods wisely [18]. Generally, classical CFR considers only user provided overall ratings for similarity computations but the crisp description of these ratings does not reflect the actual preference of human decisions. For instance, if two users are of age 14 and 17 then the distance between them would be 3, however, both of them comes under the same age group, i.e. teenager.

Fuzzification process can be used to group the users based on the distance methods. There are two ways to get fuzzy distance, first via If-Then rules and second through fuzzy distance between users. Fuzzy method chooses local distance than global distance because it gives the best for minimum distances and removes remaining. A fuzzy distance method for user u and v can be written as [11, 14].

$$fd(u, v) = \text{dis}(u, v) \times d(u, v) \tag{1}$$

where $\text{dis}(u, v)$ represents any distance method for computing distance between user u and v such as Euclidean distance formula. While, $d(u, v)$ is simply a difference operator of vectors u and v. Therefore, the fuzzy sets are designed to control systems for the decision-making process. If we want to use "age" demographic feature of a user then we cannot use it directly on the user profile because "age" feature does not reflect the actual case for human perception. Therefore, the authors in [14, 18] fuzzified "age" feature into three fuzzy sets namely, young, middle-aged, and old to get actual user group taste as shown in Fig. 2.

There is a number of proposed works related to the fuzzy set methods for RSs and the aim of the subsequent part is to give a brief overview of these works and look for some possible new perspectives. Zenebe and Norcio [24] presented a fuzzy-based method called fuzzy set theoretic method (FTM) to handles the non-stochastic uncertainty induced from vagueness, subjectivity, and imprecision in the data. The authors provided the representation methods, various fuzzy set theoretic similarity measures, and aggregation functions for calculating the recommendations ratings of items. Whereas Cao and Li [22] presented a personalized RSs for less frequently purchased items. The proposed fuzzy-based RSs retrieved optimal products based on the user's current needs collected through the user–system interactions. Researchers also have implemented a hybrid approach for fuzzy-based recommendations by combining them with different RSs techniques. A hybrid CFR-CBR conceptual framework was developed in [25] which models user preferences, item similarity, and fuzzy relations. Fuzzy set is used to represent the user preferences into two fuzzy relations, negative and positive feelings. Item similarity is computed by incorporating CBR and item-based CFR similarity. The negative and positive preferences computed as final recommendations by composing the fuzzy relations. On the other hand, hybridization of CBR and CFR was done to develop a fuzzy recommender system for rural library catalog

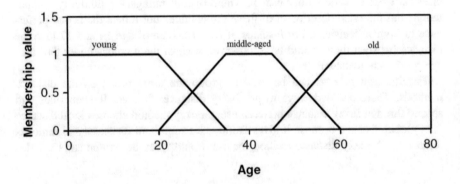

Fig. 2 Fuzzy membership function for age feature

recommendations [26]. The authors generated a fuzzy taste vector using content information to deal with the cold-start problem of collaborative recommender system.

Fuzzy sets have also been used for the development of RSs to deal with the issues of uncertainty and data sparsity in the telecom domain. Zhang et al. [27] developed a hybrid recommendations approach called FTCP-RS by combining user-based and item-based CFR through fuzzy set techniques in the telecom product/services domain. The fuzzy set techniques are designed to handle uncertainty in customer data to alleviate the data sparsity problem and improve the prediction accuracy with the help of fuzzy product similarities. A fuzzy model also has been given by researchers by combining them with other machine learning algorithms. Castellano et al. [28] developed a Neuro-Fuzzy technique for the selection of a recommender model from the usage data encoding user navigational behavior. The authors applied fuzzy clustering for generating user profile targeting the similar browsing behavior. Campos et al. [29] developed a framework by merging Bayesian network for governing the relationships between the users and fuzzy set theory for handling the uncertainty in the description of users' ratings. Similarly, the authors in [30] used a fuzzy naïve Bayesian classifier to handle the correlation-based similarity problems in CFR. A new recommender technique called reclusive methods was proposed which has two different properties than the traditional CFR [31]. First, these methods do not depend on the preferences of any user other than the preferences of the user for whom recommendations are being generated. Second, these methods use fuzzy set methods for representing and constructing heuristic rules of the items whose membership degrees are the ratings. These representations can be used to infer preferences of the user for an item. Kant and Bharadwaj [32] developed a trust and distrust-based recommender using fuzzy computational models, where fuzzy logic is used to represent both trust and distrust more naturally using linguistic expressions. Similarly, Cheng and Wang [33] represent user's subjective and objective information more naturally using fuzzy linguistic model to alleviate sparsity and the cold-start problems. While, Porcel et al. [34] developed a fuzzy linguistic-based RSs by merging CBR with the multi-granular fuzzy linguistic modeling technique, which is useful for assessing different qualitative concepts.

However, fuzzy set has shown considerable improvement in the accuracy of RSs by handling the data sparsity, cold-start, and uncertainty problems associated with the user-item features. Sometimes it is very difficult to represent the features of products, user feedback, and reasoning about their relationship in RSs because this information is subjective, imprecise and vague in nature. Developing a fuzzy user model decreases the sparsity level of the user-item matrix and improve the prediction accuracy compared to the Pearson RSs [11, 14]. In future, other fuzzy logic operators such as t-norms, implicators, aggregation operators can be used in order to check their impact on the recommendation performance. Additionally, there has been no substantial prior work of fuzzy sets in the field of context-aware recommenders and multi-criteria recommenders. Therefore, in future, it can be interesting to see how fuzzy sets can be designed for handling uncertainty in contexts and multi-criteria RSs ratings.

(b) Recommender systems based on Artificial Neural Networks

As we have discussed earlier, an artificial neural network (ANN) is inspired by the mechanism of the biological brain which has an assembly of interconnected nodes and connected links with weights. In recent years, several neural network model-based RSs have been proposed by various researchers. Such as, a feature-weighted case-based reasoning with the neural network was proposed by Kwang and Park [35]. Where the trained neural network is used to calculate feature weight set and improve retrieval accuracy of case-based reasoning. Whereas, Kano–ANN method [7] proposed to train artificial neural networks (ANN) to put users into different clusters and applied the Kano's approach to extract the implicit needs from users from different clusters.

To generate more precise movie recommendations to the users, a hybrid RSs which combines CBR with CFR was proposed by Christakou et al. [36]. Where the individual user preferences were represented through a trained ANN in the content filtering part. Similarly, Deng et al. [37] also proposed a hybrid movie recommendations approach based on ontology and neural networks. The ontology modeling gives a good method for feature extraction and captures the models and the relations among them. Whereas, the NN can get a user's preference and predict movies ratings from the user's viewpoint. The neural network was also incorporated into CFR to compute a nonlinear matrix factorization from input sparse data and side information [38]. Recurrent neural network models were also proposed to handle the time-heterogeneous feedback RSs issue [39]. The model is developed to predict the probability of a user to access a product given the time-heterogeneous feedback of this user. A session-aware RSs approach was proposed by Twardowski [40], in which the input information is obtained through user activity within a single session, denoted as a sequence of events. Now, this obtained information is incorporated into the recommendations by factorization methods and Recurrent Neural Network.

We have seen a variety of applications of ANN for the development of more accurate recommendations [4] but sometimes ANN programs become unstable when applied to solve larger problems. Also, the majority of NN are simulated on sequential machines and the operational problem occurs when attempting to simulate the parallelism of NN which gives rise to a very rapid increase in processing time with respect to the rise in the size of the problem. In future, one can come up with a solution for these problems. Additionally, one can use recurrent NN to model cross-session or long-term user goals.

(c) Recommender systems based on Evolutionary computation

Evolutionary computation is one of the techniques majorly used for the improvement of the RSs performance, which is an emerging trend with various application domains [8, 41]. The increasing interest of EC leads to the publication of the several papers in the areas of web personalization, RSs, and information retrieval. Genetic algorithms are used for solving optimization problems. There are mainly two aspects for using GA in Recommendations [3], clustering [42] and hybrid user models [14, 43]. In the following part, we will see some of the recent research suggested for GA-based RSs under clustering and hybrid used model aspects.

Kim and Ahn [42] applied GA with K-means clustering into real-world online shopping market case and compared proposed method with simple K-means algorithm and self-organizing maps. The performance of proposed GA-based K-means algorithm was demonstrated through a set of experiments compared to other typical clustering algorithms. Similarly, to generate effective recommendation a hybrid model-based movie recommendation system using improved K-means clustering and genetic algorithm was proposed, called GA-KM [20]. Principle component analysis has been employed for data reduction to dense the movie search space so that it can reduce the computation complexity for generating recommendations. Whereas, a new data clustering method has been proposed based on the concept of GA [44]. Clustering is used to create more dense clusters, which more rapidly converges in very high dimensional space. Experiments show that the proposed method alleviates the scalability issue of RSs through clustering and GA approaches. Visual-clustering recommendation (VCR) method was also proposed to solve the cold-start and sparsity problems in RSs [45]. Two different hybrid model was built between VCR and user-based (item-based) methods. The clustering of user-item was done using GA in the given model.

On the other hand, hybrid recommendation models were also proposed using GA in the past. The authors in [46, 47] presented a GA-based recommendation system in movie domain. The system used user ratings to generate movie recommendation to the target user by learning optimal weights of the features through GA. The higher weighted features are considered more significant than others. Fuzzy-genetic hybrid models were also proposed to generate more efficient movie recommendations to the users [11, 14, 48]. First, a user model was prepared by using user-item ratings, movie genre ratings, and user demographic information. The optimal weights for each user model's feature were computed through GA while computing the similarity between users using a fuzzy distance function. Silva et al. [41] combined the GA with CFR-based techniques to build a hybrid model where GA was chosen as a search algorithm and different memory-based similarity techniques chosen as examples of RSs techniques. A hybrid RSs for learning materials based on users proposed to improve the performance of recommendation [49]. Chromosomes in GA are designed using learners' implicit or latent attributes of materials then GA optimizes the weights with the help of user ratings. Final optimized weight vectors are used to generate recommendations by nearest neighborhood algorithm. Bobadilla et al. [3] proposed a GA-based metric which measures the similarity between users for movie prediction. The proposed similarity measure is used to improve the CFR technique performance in terms of MAE, coverage, precision, and recall performance quality measures by obtaining optimal similarity functions using GA.

Some people do minor changes in the simple GA to observe the change in the performance of RSs. Yilmaz and Bostanci [9] present a refined GA approach for movie recommender systems by making minor changes in recombination operation of simple GA approach. Where two individuals are selected using the roulette wheel algorithm, then in recombination phase their weights are used to generate two new offsprings by setting $\propto$ as the aggregation weight for combining alleles from both selected parent individuals. Similarly, Cui et al. [50] presented a new probabilistic genetic operator to improve the performance and effectiveness of recommendation algorithm. Interactive Genetic Algorithm (IGA) also been used to deal with the changing user preferences issue in recommendation system [15]. The initial population for IGA was generated through a fuzzy theoretic approach to CBR using reclusive methods then IGA algorithm incorporated into fuzzy CBR through user evaluation. Usually, genetic algorithm in RSs is used to learn the optimal feature weights so that similarity between two users can be computed effectively. Therefore, researchers proposed local and global similarities [51] and adaptive similarity measures [47] using GA. Whereas Alhijawi and Kilani [52] proposed a new method called SimGen, which computes the similarity among two users using the concept of GA.

Genetic Programming (GP) has also been used to alleviate sparsity of rating data and also used for improving the recommendation quality. Anand [53] employs GP for feature extraction so that a user-item space can be converted into user-feature space. This approach will reduce the data sparsity and generate a compact and dense low-dimensional preference data. Similarly, Anand [43] proposed a two-phase process for evaluating the user similarity. In the first phase, GP was employed to learn optimal transformation function which converts raw rating data to preference data, and the second phase utilizes the preference values to compute user similarity. In other work, Chong et al. [54] proposed a GP approach for CFR-based workflow recommendations, where recommendation engine is augmented by a workflow optimizer using GP through iterative evolution which leads to better workflow recommendations.

Evolutionary computing in recommendations can alleviate multiple usual problems of RSs such as data sparsity, cold-start, and new user-item by capturing optimal priorities for individual features of different users for effective similarity computation. But the major issue with GA-based recommendation is the problem of time complexity because of weight learning process. One feasible solution for such problems can be offline storage, i.e., the best set of feature weights can be stored in a separate weight matrix or on the local machine of the user. Now, these sets of weights can be used as the initial weights during future access to the system. However, one can use learnable evolution model (LEM) [55] in further research for evolving appropriate feature weights. Solution to the long-tail problem using EC also constitutes an interesting research problem to increase the diversity, novelty, and serendipity of recommendations.

(d) Recommender systems based on swarm intelligence

As we already have discussed, the idea of SI is inspired from the foraging behavior of organisms in the real world (birds, fish, and animals) to solve large-scale optimization problems. SI has been successfully applied on various fields, including different applications of data mining because of its simplicity, extendibility, and adaptability characteristics [2]. Therefore, in this subsection, we will discuss some of the recent research done on PSO-based and ACO-based recommendation techniques.

PSO is a stochastic technique to deal with optimization problems using population-based search method. Individuals in the search space are represented as particles and the movement is controlled through a set of operations. Several researchers have implemented PSO algorithms through hybrid models to overcome the limitations (computational time, accuracy) of GA in recommendations. Such as, user personal preferences were learned effectively by incorporating the PSO algorithm into the profile-matching phase of CFR-based RSs [11, 56]. Where the authors first prepared a model that comprises of three different phases. At first, different user profiles were prepared using ratings and demographic data. In the second phase, PSO algorithm incorporated to learn optimal weights of each feature of the user model and the effective similarity is computed between different user profiles. Prediction and recommendation for the target user made are in the third phase. Scalability and sparsity problems were also addressed by incorporating PSO to learn weights of various alpha estimates [57]. Here, PSO algorithm assigns different weights on the local and global neighbors based on the dependency of α value on the user and item.

Clustering techniques were also used with PSO algorithm for generating efficient recommendations. Alam et al. [58] proposed a hierarchical PSO-based clustering RSs which is based on implicit web usage data. The problem space was divided into smaller subspaces inspired from the multi-agent properties of the particles of a swarm. Each subspace represents a cluster and has a group of users, who share similar preferences. Additionally, K-nearest neighbor method was used to generate most relevant cluster for generating recommendations for the target user. Whereas data clustering problem was addressed through the concept of PSO. A hierarchical PSO algorithm was implemented by scaling the approach for better clustering [59]. While a hybrid movie RSs was proposed by classifying different types of movies according to users through PSO in order to decrease the computational complexity [60]. Where the initial parameters of PSO algorithm were obtained from K-means technique. PSO is then used to generate the initial seed to fuzzy c-means and optimizes it to obtain better results than K-means technique. A fuzzy case-based reasoning was employed in traditional collaborative filtering through PSO algorithm [61]. The PSO algorithm was incorporated in order to assign weights to features while computing the similarity between users. The effectiveness of proposed PSO-based approach was demonstrated through a set of experiments. Multi-objective PSO algorithm was also incorporated into RSs using association rule mining to improve the accuracy of CFR by handling the sparsity issue [62].

The ACO algorithm was incorporated into the trust-based RSs to improve the recommendations accuracy in the past [63–65]. Trust-based ant recommender system (TARS) [63] incorporated the concept of dynamic trust between users and selected a set of users (neighbors) based on the biological metaphor of ant colonies. Whereas, a depth-first search was performed by Semantic-enhanced Trust-based Ant Recommender System (STARS) [64] for searching the optimal trust paths in the trust network and selected the best neighbors of the target user for more accurate recommendations. ACO was also applied to several different areas such as Chen et al. [66] proposed a model for service recommendations using association rules and ACO. The relationship among the services or service sequences is extracted using association rules and ACO algorithm for recommending the navigation. An improved ant colony algorithm was proposed to optimize the micro-learning path by generating relevant recommendations of the micro-learning path according to the learner's learning position [67]. Further, ACO algorithm was also used for recommending student courses by predicting the final grades of the students [68].

The main drawback of GA-based RSs is that it takes more time for computation, since it does not keep information about the best solution in the whole process. PSO algorithm extends search by using local best solution and global best solution which is stored somewhere in the memory. Researchers have compared the GA approach with PSO approach and have shown that PSO-based RSs achieved the final solution significantly faster and more efficient compared to the GA-based RSs [11, 56]. In future, one can apply the soft computing techniques into recommendation system in a cascading manner so that common issues in the RSs can be addressed. Additionally, other weight assigning schemes can be explored in order to generate more effective recommendations.

4 Soft Computing-Based Recommender Systems Framework

There is no dedicated framework available for incorporating soft computing techniques into recommendation system. Every researcher uses a different methodology for their system development. Therefore, in this section, we will discuss a soft computing-based RSs framework which gives an idea of incorporating soft computing techniques into a collaborative filtering-based recommendation system, shown in Fig. 3. Actually, this framework has been stimulated from those researchers who have used such type of model for applying different soft computing techniques into their systems to improve the effectiveness of the recommendations [11, 14, 18, 48, 56].

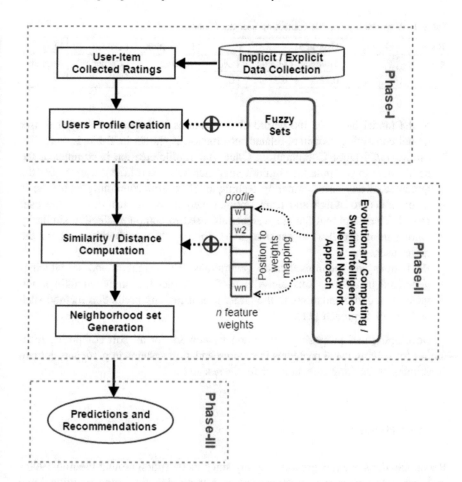

Fig. 3 Framework for incorporating soft computing techniques into recommender systems

Similar to the traditional CFR, there are three main phases shown in Fig. 3. These phases have to be accomplished in order to generate effective recommendations using any of the soft computing technique.

1. Data collection and user profile creation: First the input data is collected through different sources using the implicit or explicit technique. Explicit data can be collected by asking users' directly their preferences for the products while implicit data is collected through analyzing the behavior of the user. These data can be demographic information of the user, overall item ratings, user-item description, multi-criteria or contextual ratings given by the user. After completion of the data collection task, a user model is prepared using this collected information. A user model comprises the complete user-item information in a vector or matrix form and a fuzzy logic method can be used for the preparation

Table 1 A typical user profile containing user-item features

Rating	Gender	User age			Occupation	Genre	Mood
4	F	*Young*	*Middle*	*Old*	Teacher	Comedy	Neutral
		0.25	0.75	0			

of the model based on the nature of the features in the model. A typical user model containing user-item-related information is shown in Table 1.

2. Similarity/Distance Computation: Once user profile creation is completed, the next task is to compute the similarity/distance between different users under the guidance of any soft computing technique. Here, soft computing technique is incorporated to assign and learn optimal weights of the features for the user model. These optimal feature weights are used to compute effective similarity among users and lead to generate more appropriate neighborhood set for the target user.

3. Predictions and recommendation generations: The neighborhood set which is generated in the previous phase is used to predict the unknown ratings and generate recommendations to the target user using any prediction method such as Resnick approach [11].

Developing and proposing a common framework for all soft computing techniques for solving usual problems in recommendation systems in particular domain constitutes an exciting direction for future research.

5 Conclusion

Recommender system is growing rapidly since its inception around the mid-1990s, and researchers have started developing new methodologies using traditional and soft computing techniques concurrently. The growing use of soft computing techniques in RSs motivates us for this survey which is timely and appropriate. This paper, therefore, summarized the state-of-the-art application developments of soft computing-based recommender systems, discuss their articles into four different categories namely, fuzzy sets, neural networks, evolutionary computing, and swarm intelligence and summarizes the recently proposed techniques in each category. We included articles which highlight, analyze, and perform a study in the same area. The main feature of this survey which clearly distinguishes it from already available survey articles is that we have targeted and focused only on those papers which talk about soft computing techniques in the area of RSs. Additionally, the possible future directions of using FS, NN, EC, and SI for some of these tasks are pointed out for further research. We expect this paper will directly support researchers with the state-of-the-art knowledge of soft computing-based recommendation systems and provide a path to further develop and apply given future directions in recommender system applications.

References

1. Bobadilla J, Ortega F, Hernando A, Gu-tiérrez A (2013) Recommender systems survey. Knowl Based Syst 46:109–132
2. Grosan C, Abraham A, Chis M (2006) Swarm intelligence in data mining. Springer, Berlin Heidelberg, In Swarm Intelligence in Data Mining, pp 1–20
3. Bobadilla J, Ortega F, Hernando A, Alcalá J (2011) Improving collaborative filtering recommender system results and performance using genetic algorithms. Knowl Based Syst 24(8):1310–1316
4. Tettamanzi A, Tomassini M (2013) Soft computing: integrating evolutionary, neural, and fuzzy systems. Springer Science & Business Media, Berlin
5. Abbas A, Zhang L, Khan SU (2015) A survey on context-aware recommender systems based on computational intelligence techniques. Computing 97(7):667–690
6. Zadeh LA (1965) Fuzzy sets. Inf Control 8(3):338–353
7. Chang CC, Chen P-L, Chiu F-R, Chen Y-K (2009) Application of neural networks and Kano's method to content recommendation in web personalization. Expert Systems with Appl 36(3):5310–5316
8. Horváth T, de Carvalho AC (2016) Evolutionary computing in recommender systems: a review of recent research. Nat Comput 1–22
9. Ar Y, Bostanci E (2016) A genetic algorithm solution to the collaborative filtering problem. Expert Syst Appl 61:122–128
10. Kennedy J (2011) Particle swarm optimization. Springer, US, In Encyclopedia of machine learning, pp 760–766
11. Wasid M, Kant V (2015) A particle swarm approach to collaborative filtering based recommender systems through fuzzy features. Procedia Comput Sci 54:440–448
12. Dorigo M, Birattari M, Stutzle T (2006) Ant colony optimization. IEEE Comput Intell Mag 1(4):28–39
13. Lu J, Dianshuang W, Mao M, Wang W, Zhang G (2015) Recommender system application developments: a survey. Decis Support Syst 74:12–32
14. Al-Shamri MYH, Bharadwaj KK (2008) Fuzzy-genetic approach to recommender systems based on a novel hybrid user model. Expert Syst Appl 35(3):1386–1399
15. Kant V, Bharadwaj KK (2013) A user-oriented content based recommender system based on reclusive methods and interactive genetic algorithm. In: Proceedings of seventh international conference on bio-inspired computing: theories and applications (BIC-TA 2012). Springer, India, p 543–554
16. De Campos LM, Fernández-Luna JM, Huete JF, Rueda-Morales MA (2010) Combining content-based and collaborative recommendations: a hybrid approach based on Bayesian networks. Int J Approximate Reasoning 51(7):785–799
17. Mooney RJ, Roy L (2000) Content-based book recommending using learning for text categorization. In: Proceedings of the fifth ACM conference on Digital libraries, ACM, p 195–204
18. Wasid M, Kant V, Ali R (2016) Frequency-based similarity measure for context-aware recommender systems. Communications and Informatics, IEEE, In International Conference on Advances in Computing, pp 627–632
19. Burke R (2007) Hybrid web recommender systems. Springer, Berlin Heidelberg, In The adaptive web, pp 377–408
20. Wang Z, Xue Y, Feng N, Wang Z (2014) An improved collaborative movie recommendation system using computational intelligence. J Vis Lang Comput 25(6):667–675
21. Cai Y, Leung H-F, Li Q, Min H, Tang J, Li J (2014) Typicality based collaborative filtering recommendation. IEEE Trans Knowl Data Eng 26(3):766–779
22. Cao Y, Li Y (2007) An intelligent fuzzy-based recommendation system for consumer electronic products. Expert Syst Appl 33(1):230–240

23. Celma O (2010) Music recommendation. Springer, Berlin Heidelberg, In Music Recommendation and Discovery, pp 43–85
24. Zenebe A, Norcio AF (2009) Representation, similarity measures and aggregation methods using fuzzy sets for content-based recommender systems. Fuzzy Sets Syst 160(1):76–94
25. Cornelis C, Lu J, Guo X, Zhang G (2007) One-and-only item recommendation with fuzzy logic techniques. Inf Sci 177(22):4906–4921
26. Morawski J, Stepan T, Dick S, Miller J (2017) A fuzzy recommender system for public library catalogs. Int J Intell Syst
27. Zhang Z, Lin H, Liu K, Dianshuang W, Zhang G, Jie L (2013) A hybrid fuzzy-based personalized recommender system for telecom products/services. Inf Sci 235:117–129
28. Castellano G, Fanelli AM, Plantamura P, Torsello MA (2008) A neuro-fuzzy strategy for web personalization. In AAAI, p 1784–1785
29. de Campos LM, Fernández-Luna JM, Huete JF (2008) A collaborative recommender system based on probabilistic inference from fuzzy observations. Fuzzy Sets Syst 159(12):1554–1576
30. Kant V, Bharadwaj KK (2013) Integrating collaborative and reclusive methods for effective recommendations: a fuzzy Bayesian approach. Int J Intell Syst 28(11):1099–1123
31. Yager RR (2003) Fuzzy logic methods in recommender systems. Fuzzy Sets Syst 136(2): 133–149
32. Kant V, Bharadwaj KK (2013) Fuzzy computational models of trust and distrust for enhanced recommendations. Int J Intell Syst 28(4):332–365
33. Cheng L-C, Wang H-A (2014) A fuzzy recommender system based on the integration of subjective preferences and objective information. Appl Soft Comput 18:290–301
34. Porcel C, López-Herrera AG, Herrera-Viedma E (2009) A recommender system for research resources based on fuzzy linguistic modeling. Expert Syst Appl 36(3):5173–5183
35. Im KH, Park SC (2007) Case-based reasoning and neural network based expert system for personalization. Expert Syst Appl 32(1):77–85
36. Christakou C, Vrettos S, Stafylopatis A (2007) A hybrid movie recommender system based on neural networks. Int J Artif Intell Tools 16(05):771–792
37. Deng Y, Wu Z, Tang C, Si H, Xiong H, Chen Z (2010) A hybrid movie recommender based on ontology and neural networks. In: Proceedings of the 2010 IEEE/ACM int'l conference on green computing and communications & int'l conference on cyber, physical and social computing, IEEE computer society, p 846–851
38. Strub F, Mary J, Gaudel R (2016) Hybrid Collaborative filtering with autoencoders
39. Wu C, Wang J, Liu J, Liu W (2016) Recurrent neural network based recommendation for time heterogeneous feedback. Knowl Based Syst 109:90–103
40. Twardowski B (2016) Modelling contextual information in session-aware recommender systems with neural networks. In: Proceedings of the 10th ACM conference on recommender systems (ACM), p 273–276
41. da Silva EQ, Camilo-Junior CG, Pascoal LML, Rosa TC (2016) An evolutionary approach for combining results of recommender systems techniques based on collaborative filtering. Expert Syst Appl 53:204–218
42. Kim K, Ahn H (2008) A recommender system using GA K-means clustering in an online shopping market. Expert Syst Appl 34(2):1200–1209
43. Anand D, Bharadwaj KK (2010) Adaptive user similarity measures for recommender systems: a genetic programming approach. In: 3rd IEEE International Conference on Computer Science and Information Technology (ICCSIT), vol. 8. IEEE, p 121–125
44. Georgiou O, Tsapatsoulis N (2010) Improving the scalability of recommender systems by clustering using genetic algorithms. In: International conference on artificial neural networks. Springer, Berlin, Heidelberg, p 442–449
45. Marung U, Theera-Umpon N, Auephanwiriyakul S (2016) Top-N recommender systems using genetic algorithm-based visual-clustering methods. Symmetry 8(7):54

46. Fong, Simon, Yvonne Ho, and Yang Hang (2008) Using genetic algorithm for hybrid modes of collaborative filtering in online recommenders. In: Eighth international conference on hybrid intelligent systems, HIS'08, IEEE, p 174–179

47. Anand D, Bharadwaj KK (2010) Enhancing accuracy of recommender system through adaptive similarity measures based on hybrid features. In: Asian conference on intelligent information and database systems. Springer, Berlin, Heidelberg, p 1–10

48. Ujjin S, Bentley PJ (2002) Learning user preferences using evolution. In: Proceedings of the 4th Asia-Pacific Conference on Simulated Evolution and Learning, Singapore

49. Salehi M, Pourzaferani M, Razavi SA (2013) Hybrid attribute-based recommender system for learning material using genetic algorithm and a multidimensional information model. Egypt Inf J 14(1):67–78

50. Cui L, Ou P, Fu X, Wen Z, Lu N (2016) A novel multi-objective evolutionary algorithm for recommendation systems. J Parallel Distrib Comput

51. Anand D, Bharadwaj KK (2011) Utilizing various sparsity measures for enhancing accuracy of collaborative recommender systems based on local and global similarities. Expert Syst Appl 38(5):5101–5109

52. Alhijawi B, Yousef K (2016) Using genetic algorithms for measuring the similarity values between users in collaborative filtering recommender systems", In: IEEE/ACIS 15th International Conference on Computer and Information Science (ICIS), IEEE, p 1–6

53. Anand D (2012) Feature extraction for collaborative filtering: a genetic programming approach. Int J Comput Sci Issues 9(1)

54. Chong CS, Zhang T, Lee KK, Hung GG, Lee B-S (2013) Collaborative analytics with genetic programming for workflow recommendation. In: IEEE International conference on systems, man, and cybernetics (SMC), IEEE, p 657–662

55. Michalski RS (2000) Learnable evolution model: Evolutionary processes guided by machine learning. Mach Learn 38(1–2):9–40

56. Ujjin S, Bentley PJ (2003) Particle swarm optimization recommender system. In: Proceedings of the 2003 Swarm Intelligence Symposium (SIS'03, IEEE), p 124–131

57. Bakshi S, Jagadev AK, Dehuri S, Wang GN (2014) Enhancing scalability and accuracy of recommendation systems using unsupervised learning and particle swarm optimization. Appl Soft Comput 15:21–29

58. Alam S, Dobbie G, Koh YS, Riddle P (2014) Web usage mining based recommender systems using implicit heterogeneous data. Web Intell Agent Syst Int J 12(4):389–409

59. Alam, S, Dobbie G, Riddle P (2011) Towards recommender system using particle swarm optimization based web usage clustering. In: Pacific-Asia conference on knowledge discovery and data mining, Springer, Berlin, Heidelberg, p 316–326

60. Katarya R, Verma OP (2016) A collaborative recommender system enhanced with particle swarm optimization technique. Multimedia Tools Appl 75(15):9225–9239

61. Tyagi S, Bharadwaj KK (2014) A particle swarm optimization approach to fuzzy case-based reasoning in the framework of collaborative filter-ing. Int J Rough Sets Data Anal (IJRSDA) 1(1):48–64

62. Tyagi S, Bharadwaj KK (2013) Enhancing collaborative filtering recommendations by utilizing multi-objective particle swarm optimization embedded association rule mining. Swarm Evol Comput 13:1–12

63. Bedi P, Sharma R (2012) Trust based recommender system using ant colony for trust computation. Expert Syst Appl 39(1):1183–1190

64. Gohari FS, Haghighi H, Aliee FS (2016) A semantic-enhanced trust based recommender system using ant colony optimization. Appl Intell 1–37

65. Tengkiattrakul P, Maneeroj S, Takasu A (2016) Applying ant-colony concepts to trust-based recommender systems. In: Proceedings of the 18th International Conference on Information Integration and Web-based Applications and Services (ACM), p 34–41

66. Chen Z, Shao Z, Xie Z, Huang X (2010) An attribute-based scheme for service recommendation using association rules and ant colony algorithm. In: Wireless Telecommunications Symposium (WTS) IEEE, p 1–6
67. Zhao Q, Zhang Y, Chen J (2016) An improved ant colony optimization algorithm for recommendation of micro-learning path. In: IEEE international conference on computer and information technology (CIT) IEEE, p 190–196
68. Sobecki J, Tomczak JM (2010) Student courses recommendation using ant colony optimization. In: Asian conference on intelligent information and database systems. Springer, Berlin, Heidelberg, p 124–133

Part II
Soft Computing Based Online Documents Summarization

Hierarchical Summarization of News Tweets with Twitter-LDA

Nadeem Akhtar

1 Introduction

Twitter is more used social network for grasping news than any other social network. According to a survey conducted by the American Press Institute funded by Twitter in 2015 [1], around 86% of the Twitter users in America use Twitter for news and 74% of them use it daily. Twitter is mostly used for breaking news alerts (40%) and finding news in general (39%) than keeping social relations by communicating what I am doing or thinking about or keeping in touch with the people I know. 28% of the users also use Twitter to know about a live event, i.e., a sports event, a concert. In some instances, Twitter is faster to spread news than traditional news media in emergency situations like earthquake [2].

Several works have been done to extract news from Twitter data or stream. A newsworthiness model [3] trains a classifier through active learning using content and journalism-based features and finds most relevant and newsworthiness tweets. SUMBLR [4] system proposes continuous summarization of evolving tweet stream to analyze dynamic, real-time, and large-scale stream. Several works [5–8] have focussed on analyzing an event-related tweet stream to monitor the evolution of event and generate storylines.

Some works have also discovered connection and correlation between online news media document and Twitter news. EKNOT [9] system identifies major events in online news documents and connects it to tweets to provide a comprehensive summary of both news and social media. SociRank [10] provides an unsupervised framework which extracts important news from news media and Twitter and ranks them by relevance according to their media focus, user attention, and user intention.

N. Akhtar (✉)
Department of Computer Engineering, Aligarh Muslim University,
Aligarh 202001, India
e-mail: nadeemakhtar@zhect.ac.in

© Springer Nature Singapore Pte Ltd. 2017
R. Ali and M. M. S. Beg (eds.), *Applications of Soft Computing for the Web*,
https://doi.org/10.1007/978-981-10-7098-3_6

The news provided by the traditional news sources is verified by the journalists and they are held accountable for their content. Whereas, anyone can publish any unverified news item on Twitter, which may be wrong or useless. News items on Twitter must be checked for relevance and may be ranked according to their information content. Because of the unverified, mostly irrelevant and noisy nature of general Twitter news, most of the Twitter users tend to follow known popular journalists, writers, commentators, and news media Twitter handles for verified, relevant and informative news content [1]. These followers interact socially by either retweeting or responding to the breaking news by posting their comments and opinions. The responses to some breaking news may start a long social discussion on Twitter. Though the number of tweets generated by a news media handle may be small, i.e., a few hundred in a single day, the number of retweets and replies may be very large spanning several days. To provide a topic-oriented concise news summary of these news-related tweets, a summarization framework is needed that can automatically identify individual topics/events and generate a concise and cohesive tweet summary of the topic/event.

In this work, a news summarization system is developed to summarize multi-topic tweets data of verified Twitter news sources and their social interaction with users. Since only verified Twitter news sources, i.e., news media, journalists, commentator's handles, and their social interactions are considered, a summary of the proposed summarization framework will be reliable and verified.

In the proposed framework, a topic model is used to identify the major topics/events in the data. For each identified topic, all topical tweets are extracted and mined for identifying important tweets based on their content and social interaction. The optimal size of tweet summary is obtained by using an optimization algorithm. The tweet summary may be further explored by applying the optimization algorithm recursively to get detailed tweet summary for getting a sub-topic/sub-event of the news.

Rest of the paper is organized as follows. Next section discusses related work. The section following related works presents the proposed system. After that experimental settings and results are discussed followed by the evaluation and discussion. Finally, conclusion and future work are presented.

2 Related Work

The proposed framework presents a topic/event-oriented tweet summary of the Twitter news and its social interaction in an evolving hierarchy. This work is related to topic modeling, event detection in Twitter, hierarchy construction and joint study of news and social network. These related works are briefly reviewed next.

2.1 Topic Modeling

Generating topic-oriented tweet summary is similar to storyline generation, which detects and tracks topics in temporal data. Storylines are generated using topic detection and tracking (TDT) [11, 12] which aims to link streams of texts. TDT has five major components: story segmentation, topic tracking, topic detection, first story detection, and link detection.

Another method is finding the latent topics in data using a topic model and then generating a temporal summary. Topic models find latent topics and themes in data by considering word co-occurrence. There has been a lot of research and a variety of topic models have been proposed—simple LDA topic model [13], dynamic topic model [14], hierarchical topic model [15], Twitter-LDA topic model [16]. The topic model usually does not provide good results for short texts like tweets. Twitter-LDA, a variation of Latent Dirichlet Allocation, which assumes a single topic for each tweet and pools all the tweets of a user into a single document, provides better results. In this work, Twitter-LDA is tailored to the needs of news summary generation in Twitter. Instead of pooling the tweets of users, experiments are performed on pooled documents formed by pooling on the basis of replies and hashtags.

2.2 Event Detection

Many works have focused on mining trending events on Twitter. Most works [17–19] detect peaks in the temporal tweet stream. These works measure the change in the number of tweets, word distribution etc. in successive time intervals to identify peaks in the tweet stream. In [20], both user's interest and temporal information are integrated for bursty event detection. The proposed work also considers the pattern of social interaction of users to find important events besides considering changes in the word distribution. Social interaction of the users is measured by retweets and replies, which is similar to [21], which uses only retweets for political learning.

Tweet summarization is also coupled with event detection. GEAM [22] associates topics and event aspects, i.e., time, location, entities, keywords to the identified events. EventSense [23] also models sentiments of the users besides identifying the events. In [24], a phrase reinforcement graph method is used to find important phrases and related events and summary tweets.

Some event detection systems [25, 26] provides hierarchical summarization, whose summary can be rolled up/down to give less/more summary according to user's interests. In the proposed work, a similar hierarchical summary is generated on demand by applying the recursive procedure in the selected time interval.

2.3   Joint Study of News and Social Network

Because of several potential applications, a joint study of news and social media has got attention in recent research. In [27], the authors performed a comparative study between news and tweets by applying separate topic models. Gao et al. [28] developed a joint topic model for event summarization across news and social media. Wei and Gao [29] studies news highlight extraction from tweets for a given event.

In all of the works focused on the joint study of news and social media, a separate news source is used to compare or jointly model with Twitter. In the proposed work, no external news source is used. Only the traditional news media and popular journalists' Twitter handles are used. This provides a restricted analysis of Twitter news and its social interaction.

3   Proposed Method

The overview for the proposed Twitter News Summarization framework is depicted in Fig. 1. There are five key stages: Tweet Preprocessing, Tweet Pooling, News Topic Detection, Tweet Selection, and Hierarchical Tweet Summary on Timeline. In the following subsections, each of these stages is presented.

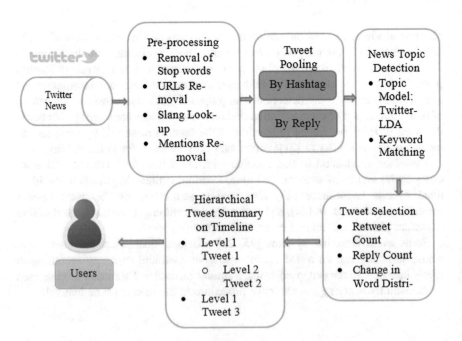

Fig. 1 Overview of the proposed system

3.1   Tweet Preprocessing

First, the data collected from Twitter are processed to remove the quirks of the social nature of Twitter. A raw tweet may have any sort of slangs, punctuations, and hyphenation. After tokenizing, all stop-words, expressions, mentions, and URLs are removed in the Tweet text. HTML characters are escaped and slangs are modified using a slang dictionary. Only those tweets are retained which has more than 3 tokens after preprocessing.

3.2   Tweet Pooling

Tweets are short texts, containing a maximum of 140 characters. Training topic models directly on individual tweet data do not provide good results [30]. The solution to this problem is to pool or merge similar tweets into a single document, then apply topic models to the pooled tweets. Two main tweet pooling techniques [30] are: pooling by the same user, pooling by the same hashtag. Recently, a new pooling technique is used to merge tweets on the basis of conversation [31]. Every tweet has an in_reply_to_status_id field, which has value 0 if the tweet is not posted in reply to some previous tweet. If a tweet A is posted in reply to some previous tweet B, in_reply_to_status_id field of tweet A is set to the tweet_id of the previous tweet B. Some other tweet C may be posted which may be in reply to either tweet A or tweet B. Since these tweets are in the same conversation, they are semantically related and belong to same topics. All the tweets in such conversations are merged to make longer documents.

3.3   News Topic Detection

Since similar tweets have been pooled, a variation of standard LDA [13] is used to identify the important news topics in the tweet set. Due to the short size of tweets, a simplifying assumption is made that each tweet of the pooled document has a single topic like in Twitter-LDA [16]. The generation process of tweets is shown in Fig. 2.

There are T topics in the tweet set. Φ^t and Φ^β are word distributions for topic t and background words respectively. θ^d is the topic distribution for pooled document d. π is a Bernoulli distribution that decides between topical word and background word. To generate a pooled document, first, a topic distribution is chosen. Then for each tweet of the pooled document, a single topic is drawn. After that, individual words are drawn either from the topic distribution or from background word distribution based on Bernoulli distribution π.

1. Draw $\Phi^\beta \sim \text{Dir}(\beta)$, $\pi \sim \text{Dir}(\gamma)$
2. For Each News Topic T
    Draw $\Phi^t \sim \text{Dir}(\beta)$,
3. For each pooled document D
    Draw $\theta^d \sim \text{Dir}(\alpha)$,
4. For each tweet m in the pooled document
    i. $z_{d,m} \sim \text{Multi}(\theta^d)$
    ii. For each word n in the tweet
        Draw $y_{d,m,n} \sim \text{Multi}(\pi)$
        If $y_{d,m,n} = 0$ Draw $w_{d,m,n} \sim \text{Multi}(\Phi^\beta)$
        If $y_{d,m,n} = 1$ Draw $w_{d,m,n} \sim \text{Multi}(\Phi^{Z_{d,m}})$

Fig. 2 Generation process of tweets

3.4 Tweet Selection

A user of the proposed system may view the summarized news of entire tweet dataset or she may provide some keywords to find the summary of keywords-related news. In the later case, simple keyword matching between user-provided keywords and top words of each topic is used to select the most relevant topic.

All the tweets related to each topic are found. For this, the similarity between the tweets and topics are found using keyword matching. Each tweet is assigned to the topic for which at least one tweet word is common with topic keywords. This simple keyword matching scheme can be relied upon because we have selected only the top words of a topic for keyword matching. After this step, the original tweet dataset is segmented into several segments, one for each selected topic.

Each segment has a large number of tweets. Some are simply retweets or have almost the same content to other tweets. The summary of each topic should present the most important cohesive tweets without duplication. Moreover, the tweets should be arranged on a timeline.

To find the summary tweets for a topic, an optimization algorithm is used which optimizes the social interactions and textual contents of the tweets. First, all the tweets in the topic segment are arranged according to their timestamp. There may be a large number of tweets in a topic segment which may sharply increase computation time if processing is done on individual tweets. Also since individual tweets are very short, it is difficult to measure the change in the textual content of the tweets. To resolve these problems, binning is used to aggregate tweets on the basis of time interval, assuming there be little change in the topic of adjacent tweets. All the tweets in the time interval I are binned into the same bin. The value of time interval I is kept small and chosen empirically for the dataset. Now, the problem of finding best summary tweets for a topic is transformed into finding the best bins that contain best summary tweets in the topic segment after binning. Next, two types of measures which are used to select the summary tweets are described.

- Social Interaction of the tweet: This measures the social aspect of the tweet. There are several auxiliary metadata such as tweet mentions, URLs, retweet, and reply count which accounts for the social interaction among the tweets. In this work, we have considered only retweet and reply count to keep the system simple. A breaking news tweet posted by a news media is retweeted several times for sharing. Also, several users post their comments on an important and interesting news tweet several times. A news tweet is important if either it is retweeted or replied several times. The retweet count and reply count is found for each tweet and summed up. The social aspect is calculated as follows:

$$\text{Social Interaction} = \log_{2 * max}(\text{RR}) \tag{1}$$

where RR = Retweet_Count + Reply_Count for all the tweets in the bin. max is the maximum value of RR for any tweet.

- Textual Content of the tweet: All the tweets are related to a single topic but composed of several subtopics in a segment and are arranged on a timeline. Identifying important tweets in the segment mean identifying important subtopics. Any subtopic change in segment tweets will cause a significant change in the distribution of words of the tweets. For each bin in the topic segment, simple LDA is used to find the word probability distribution. To quantify the change in word distributions, Hellinger distance is used to measure the difference between two consecutive bins. Hellinger distance between two probability distributions P_a and P_b of vocabulary length N each is given by the following formula:

$$\text{Change}_{\text{Textual Content}} = \sum_{v=1}^{N} \left(\sqrt{P_{a,v}} - \sqrt{P_{b,v}} \right)^2 \tag{2}$$

- To find the best number of bins, weighted average of semantic and social aspects are optimized using Gap statistics optimization method [32]. The optimization criterion is defined as

$$\text{Maximize } W_k = \alpha \times \text{social interaction} + (1 - \alpha) \times \text{Textual Content} \tag{3}$$

Here, $0 < \alpha < 1$ is weighting constant, which controls the effect of social and semantic aspects in optimization criterion. Constant k is the number of bins in the topic segment. To determine the best value for k Gap statistics method is used, which determines the appropriate number of clusters k in a clustering problem by maximizing the gap between expectation of data sample of size n and a corresponding reference distribution.

$$\text{Gap} = E_n^*\{\log(W_k)\} - \log(W_k) \tag{4}$$

where W_k is score of topic segment with k best bins. Reference distribution is generated uniformly over the observed features.

- Once best bins are found, representative tweet(s) are selected in each of the best bin and arranged on a timeline according to their timestamp. The representative tweet in the bin is that for which the social interaction and textual content measures are maximum. Here, a textual content measure of a tweet is found by comparing the tweet probability distribution t with the bin probability distribution b using Hellinger distance.

$$\text{Representative\_Tweet}_b = \arg\max(\text{Retweet\_Count}_t + \text{Reply\_Count}_t$$
$$+ \sum_{v=1}^{N} \left(\sqrt{p_{t,v}} - \sqrt{p_{b,v}}\right)^2 \tag{5}$$

4 Experiments and Results

4.1 Dataset

To evaluate the efficacy of the proposed system, BBC news tweets are collected using Twitter streaming API [33]. All the tweets containing the term 'bbc' or tweeted by the official handle of BBC News are collected from March 22, 2017 to April 02, 2017. Some tweets in the collected dataset have replied to the tweets that were posted before March 22, 2017. Such tweets were identified and also collected to complete the dataset. The total number of tweets in the dataset is 152,936.

4.2 Parameter Setting

The gap statistic parameter B is set to 10. Time interval I for different topics is taken differently. For large tweet-density topics, I is kept small and for smaller tweet-density topics, I is kept large. For example, time interval I for UK parliament attack topic for level 1 and 2 is 60 and 20 respectively. For Brexit topic, time interval I at level 1 and 2 is 30 and 10 respectively.

The number of levels in the tweet summary hierarchy is set to 2. The number of keywords per topic is set to 30. The value for the weighting parameter α is taken as—0, 0.5 and 1.0.

4.3   Evaluation Methodology

The system is divided into two parts for the purpose of evaluation. In the first part, news topics detection is evaluated and in the second part, tweet summaries are evaluated. Since text summaries are intended to be used by a human, human evaluation method is used.

4.4   News Topic Detection Evaluation

Manual labeling is used for labeling the topics. The topics found are analyzed and a label is assigned manually to each topic that is most similar to the topical words. To evaluate the precision of words in each topic, the percentage of relevant keywords is considered. Only top 30 words are considered for finding the precision for each topic. Manual labels and keywords precision are found for both the tweet pooling strategies—by hashtags and replies.

4.5   Tweet Summary Evaluation

The summary tweets are evaluated for the topic coherence using precision and recall evaluation metric. For each of the two selected topics, 15 important event tweets are identified from two internet sources—Google News and Wikipedia to construct the baseline events [34, 35]. For each of the selected algorithm with which the proposed approach is compared, top 25 tweets are considered to find precision. To find precision for the proposed approach, first 25 tweets at the first level are considered. If the number of tweets at the first level is less than 25, second-level tweets are also considered to make it equal to 25. Topic precision is the percentage of relevant tweets in the top 25 tweet summary. Topic recall is the percentage of summary tweets, which are same as one of the 15 important event tweets. Some tweets having different texts in the summary are similar to one another pointing to the same event. Such tweets are counted as one in the calculation of precision and recall.

The results are reported for three different values of α-
0.0: only textual content is considered
1.0: only social interaction is considered
0.5: both social interaction and textual content are considered equally.

4.6   Results

In Table 1, top 10 keywords and manually assigned labels are shown for each of the top 5 topics, when pooling is done based on hashtags. In Table 2, results are shown when pooling is done based on replies. Both strategies result the same top three topics—UK parliament attack, Scotland referendum, The Premier League Show. The number of relevant topics found in hashtag pooling is comparatively much smaller than in replies pooling. Also, the number of relevant keywords for a topic is much smaller in hashtag pooling than in replies pooling.

Due to space restrictions, all tweet summary results cannot be shown. Tweet summary results are shown for only two topics—'UK Parliament Attack' found using topic modeling and 'Brexit' using user provided keyword. Some tweets in the tweet summary results for 'UK parliament attack' and 'Brexit Exit Vote' events are shown in Figs. 3 and 4 respectively. The value for parameter α is 0.5 and 1.0 for Brexit and UK parliament event respectively.

Table 1 Top 10 keywords for hashtags pooling

Manual label	Top 10 keywords	Keywords precision
UK Parliament Attack	Westminster terror Brexit battle round classic sets prank stories report	16.67
Scotland Referendum	Scotland vote letter referendums independence ago people million support neutral	30.0
Premier League Show	Live premier league season left minutes leicester fans wins record	43.34
Mosul	Heartbreaking tears fighting found back watch city death flee conflict	36.67
Misc	Mail daily man saido music faced statement ban drugs led	–

Table 2 Top 10 keywords for replies pooling

Manual label	Top 10 keywords	Keywords precision
UK Parliament Attack	Police islam muslim terrorist attack London isis dead shot victims	96.67
Scotland Referendum	Scotland people british bye Ireland referendum voted country vote biased	56.67
The Premier League Show	League football team cup FIFA Messi game arsenal season win	93.34
Trump	Money work Trump law gold jobs Obama wall Mexico administration	60.0
Tennis	Tennis beat cheat team Johanna final player sport official Federer	36.67

3/22/17 12:47 Brexit MPs?Plenty think the other view that BBC too supportive of Brexit. Possibly balance is right then? No one happy.

3/25/17 3:26 Brexit is "a tragedy and a failure" - @JunckerEU's exclusive BBC interview as UK prepares to trigger EU exit process

> 3/27/17 9:56 AM, theresa_may says #brexit is an opportunity to "build a more united nation" and make Britain an "unstoppable force

3/27/17 1:32 PM, PM says Brexit-plan to 'strengthen Union' ahead of Sturgeon meeting

3/28/17 10:47 Brexit: Theresa May eyes 'special partnership' with EU

> 3/30/17 10:56 AM, Brexit Secretary David Davis: Great Repeal Bill will provide no future role for the European Court of Justice

> 3/30/17 11:24 AM, Brexit: No talks with the UK before clear withdrawal terms, says France

> 3/30/17 12:21 AM, Brexit sovereignty plan set out in Great Repeal Bill

Fig. 3 Tweet summary of Brexit Event

Two levels of the hierarchy are shown in the tweet summary. Second-level summary tweets are more indented.

The results are compared with three summarization algorithms—term frequency-based algorithm TFIDF, probability-based algorithm SumBasic [36], and centroid-based algorithm SUMMA [37]. Since retweets and replies are considered in the proposed approach, which measures the frequency of words in some way, TFIDF and SumBasic algorithms are chosen for comparison. Only tweet texts are utilized to produce summary using these three algorithms. In Table 3, precision and recall for the three selected events and the proposed approach are shown. Event 'UK parliament attack' is taken from topic modeling results and the tweets for 'Brexit' event is taken by user-specified keyword searching. The results are shown for the three values for α— 0, 0.5, and 1.0 for the proposed approach.

5 Discussion

5.1 News Topic Detection

A topic model finds general topics by considering word co-occurrence relationship. More often, it fails to identify events because a news event is not only characterized by word distribution but also by time, named entities like person, location [38]. The results show the proposed approach is able to find coherent word

3/22/17 11:39 AM, Westminster latest:-Police officer stabbed at parliament, suspect shot-Car drives into people on bridge-PM safe

3/22/17 10:39 AM, police *fined* her for supposedly "wasting their time" when she reported his violent behaviour. #WhyWomenDontReport

3/22/17 12:18 PM, Police say attack at UK Parliament being treated as "terrorist incident until we know otherwise

3/22/17 2:20 PM, 4 people have died in an attack in London. This video shows the moment the incident began outside the UK Parliament

3/23/17 3:01 AM, Thoughts of PM with those killed, injured and their families in "appalling" Westminster terror attack, says No

3/23/17 4:03 AM, Police revise #Westminster attack death toll to 4: woman in her 40s, man in his 50s police officer attacker

3/25/17 4:15 AM, Islam is tolerant and peaceful religion someone have said

3/25/17 5:37 AM, Six people arrested as part of investigation into Westminster terror attack have been released without charge

3/27/17 12:43 PM, The terrorists will not defeat us" says UK home secretary at Trafalgar Square vigil for London attack victims

3/28/17 6:29 AM, Police investigating 'human waste in Coca Cola cans

3/28/17 8:10 AM, The attack wasn't related to racism but drugs. There was news of death of a youth due to forced overdose by Nigerians

3/29/17 10:50 AM, People link hands and fall silent on Westminster Bridge to mark a week since the London attack

3/30/17 11:08 PM, so how many people have jumped the shark?

Fig. 4 UK Parliament Attack tweet summary

Table 3 Precision and recall for the summarization algorithms

S no	Event	UK Parliament Attack			Brexit		
		Alpha	Precision	Recall	Alpha	Precision	Recall
1	TFIDF	–	0.48	0.60	–	0.36	0.60
2	SumBasic	–	0.40	0.66	–	0.28	0.53
3	SUMMA	–	0.32	0.40	–	0.20	0.26
4	Proposed approach	0.0	0.52	0.53	0.0	0.44	0.55
		0.5	0.54	0.46	0.5	0.43	0.46
		1.0	0.48	0.67	1.0	0.39	0.60

distributions that clearly indicate a single major event when the tweet pooling is done on the basis of replies. Whereas, hashtags-based tweet pooling does not result in good events.

The reason for this is that users tend to reply to a news event using in_reply_to_status_id field instead of using tweet mentions or hashtags. A Twitter news user mostly retweets the breaking news tweet in its original form or replies to it using in_reply_to_status_id field. The Twitter user does not usually use tweet mentions or hashtags while retweeting or responding to a breaking news. Only 15 and 26% of the tweets in the collected news tweet dataset contains a hashtag and tweet mention respectively. A few of the hashtags and tweet mentions belong to a large number of tweets belonging to a few number of events.

5.2 Tweet Summary

Only two levels are shown in the tweet summary results. But as many levels of the summary can be generated as desired by the user in the selected time period. Since exactly the same process is used recursively with different parameter values to find more important news tweets at lower levels, tweets at the lower levels may not be logically related to corresponding higher level tweets with parent–child coherence, but they are also important tweets; slightly less important than the tweets at higher levels of hierarchy.

5.2.1 Comparison with Other Algorithms

In Table 3, the results obtained for the proposed approach with $\alpha = 1.0$ (only social interaction) is comparable with TFIDF algorithm. The proposed approach with $\alpha = 1.0$ use retweets and replies to find best tweets which are similar to using term frequency for finding important tweets in TFIDF.

Since important 15 events chosen for each topic mostly includes highly retweeted tweets, recall for TFIDF and SumBasic is good but less than the proposed approach. The proposed approach also considers tweet importance based on conversational relationship and topic terms co-occurrence.

The precision of the proposed approach is better than other two algorithms— SumBasic and SUMMA. SumBasic uses term probabilities to find important tweets. It reduces the term probabilities of the tweets once they are included in the summary. This denies inclusion of important tweets having terms with reduced probabilities but related to several important events. SUMMA fails to capture the importance of retweets and replies of the tweeting behavior.

5.2.2 Effect of α

Precision values for the selected topics are less when only social interaction ($\alpha = 1.0$) is used. The reason for this is that two or more different tweets about the same event may be retweeted a large number of times. These tweets are found in the summary but counted as one because of the event similarity.

Recall values for the selected events are better when the weightage of textual content is more. The reason for this is that Twitter news users retweet or reply to the main breaking news more than the other less important news. The main news is easily found while considering the retweet and reply count. But other less important news tweets cannot be found using only retweet and reply count.

Recall of the system is increased when more levels of tweet summary are generated but it also results in decreased precision.

Precision value for the user-specified keywords based topics, i.e., Brexit is large because all the retrieved tweets belong to such topic. But recall value for these topics is low because some important tweets that do not contain user-specified keywords are not considered.

6   Conclusion

An extractive tweet summarization system is proposed and implemented for Twitter news users. A modified Twitter-LDA topic model is utilized with tweet pooling based on hashtags and replies to find main topics in the tweet dataset. For each topic, tweet summary is generated by finding important tweets using retweet and reply count and change in word distributions along the temporal dimension. Tweet summary can be generated hierarchically by applying the process recursively. The results have shown that good extractive summary can be produced with satisfactory precision and recall values. Topic models are able to identify general news topics encompassing several news events. But it fails to identify individual news events.

Future works include considering named entities and temporal information to better identify the news topics. Besides using the retweet and reply count, other auxiliary information present in the Twitter such as mentions, hyperlinks, location, and multimedia data can also be incorporated to improve the summary results.

References

1. Rosenstiel T et al (2015) Twitter and the news: how people use the social network to learn about the world. American Press Institute, Arlington
2. Shi B, Georgiana I, Neil H (2014) Insight4news: Connecting news to relevant social conversations. In: Joint European conference on machine learning and knowledge discovery in databases. Springer, Berlin
3. Ross J, Krishnaprasad T (20156) Features for ranking tweets based on credibility and newsworthiness. In: Collaboration technologies and systems (CTS), 2016 international conference on. IEEE, 2016
4. Shou L et al (2013) Sumblr: continuous summarization of evolving tweet streams. In: Proceedings of the 36th international ACM SIGIR conference on research and development in information retrieval. ACM, 2013
5. Lin C et al (2012) Generating event storylines from microblogs. In: Proceedings of the 21st ACM international conference on Information and knowledge management. ACM, 2012
6. Chakrabarti Deepayan, Punera Kunal (2011) Event Summarization Using Tweets. ICWSM 11:66–73

7. Nichols J, Jalal M, Clemens D (2012) Summarizing sporting events using twitter. In: Proceedings of the 2012 ACM international conference on intelligent user interfaces. ACM, 2012

8. Shen C et al (2012) A participant-based approach for event summarization using twitter streams. In: HLT-NAACL

9. Li M et al (2016) EKNOT: event knowledge from news and opinions in Twitter. In: Proceedings of the thirtieth AAAI conference on artificial intelligence. AAAI Press

10. Davis D, Figueroa G, Chen Y-S (2016) SociRank: identifying and ranking prevalent news topics using social media factors. In: IEEE transactions on systems, man, and cybernetics: systems

11. Allan, James, ed. Topic detection and tracking: event-based information organization. Vol. 12. Springer Science & Business Media, 2012

12. Allan J (ed) (2012) Topic detection and tracking: event-based information organization, vol 12. Springer Science & Business Media, Berlin

13. Blei DM, Ng AY, Jordan MI (2003) Latent dirichlet allocation. J Mach Learn Res 3:993–1022

14. Blei DM, Lafferty JD (2006) Dynamic topic models. In: Proceedings of the 23rd international conference on machine learning. ACM

15. Griffiths DMBTL, Tenenbaum MIJJB (2004) Hierarchical topic models and the nested chinese restaurant process. Adv Neural Inf Process Syst 16:17

16. Zhao WX et al (2011) Comparing twitter and traditional media using topic models. In: European conference on information retrieval. Springer, Berlin

17. Abdelhaq Hamed, Sengstock Christian, Gertz Michael (2013) Eventweet: online localized event detection from twitter. Proceedings of the VLDB Endowment 6(12):1326–1329

18. Akhtar N, Siddique B (2016) On hierarchical visualization of event detection in twitter. In: IC4S, Ajmer, 12–13 Aug 2016

19. Siddique B, Akhtar N (2017) Temporal hierarchical event detection of timestamped data. IEEE ICCCA—2017, Noida, 05–06 May 2017

20. Diao Q, Jiang J, Zhu F, Lim E-P (2012) Finding bursty topics from microblogs. In: Proceedings of the 50th annual meeting of the association for computational linguistics (ACL), p 536–544

21. Wong FMF et al (2016) Quantifying political leaning from tweets, retweets, and retweeters. IEEE transactions on knowledge and data engineering 28(8):2158–2172

22. You Y et al (2013) GEAM: a general and event-related aspects model for twitter event detection. WISE 2013

23. Schinas E et al (2013) Eventsense: capturing the pulse of large-scale events by mining social media streams. In: Proceedings of the 17th Panhellenic conference on informatics. ACM, 2013

24. Sharifi B, Hutton M-A, Kalita J (2010) Summarizing microblogs automatically. Human language technologies: the 2010 annual conference of the North American chapter of the association for computational linguistics. Association for computational linguistics

25. Otterbacher J, Radev D, Kareem O (2006) News to go: hierarchical text summarization for mobile devices. In: Proceedings of the 29th annual international ACM SIGIR conference on research and development in information retrieval. ACM, 2006

26. Christensen J, Stephen Soderland M, Etzioni O (2013) Towards coherent multi-document summarization. In: Proceedings of NAACL 2013

27. Zhao WX, Jiang J, Weng J, He J, Lim E-P, Yan H, Li X (2011) Comparing twitter and traditional media using topic models. ECIR. Springer, Berlin, pp 338–349

28. Gao W, Li P, Darwish K (2012) "Joint topic modeling for event summarization across news and social media streams", in CIKM. ACM, New York, NY, USA, pp 1173–1182

29. Wei Z, Gao W (2014) Utilizing microblogs for automatic news highlights extraction. In COLING, p 872–883

30. Mehrotra R et al (2013) Improving lda topic models for microblogs via tweet pooling and automatic labeling. In: Proceedings of the 36th international ACM SIGIR conference on research and development in information retrieval. ACM, 2013
31. Alvarez-Melis D, Martin S (2016) Topic modeling in twitter: aggregating tweets by conversations. In: ICWSM
32. Tibshirani Robert, Walther Guenther, Hastie Trevor (2001) Estimating the number of clusters in a data set via the gap statistic. J Royal Stat Soc Ser B Stat Methodol 63(2):411–423
33. Makice K (2009) Twitter API: up and running: learn how to build applications with the twitter API. O'Reilly Media, Inc.
34. https://en.wikipedia.org/wiki/2017_Westminster_attack. Accessed on 10th Apr 2017
35. https://en.wikipedia.org/wiki/Brexit. Accessed on 10th Apr 2017
36. Nenkova A, Lucy V (2005) The impact of frequency on summarization. Microsoft research, Redmond, Washington, Tech. Rep. MSR-TR-2005 101
37. Saggion H (2014) Creating summarization systems with SUMMA. In LREC, p 4157–4163
38. Marcus A et al (2011) Twitinfo: aggregating and visualizing microblogs for event exploration. In: Proceedings of the SIGCHI conference on human factors in computing systems. ACM, 2011

Part III
Soft Computing Based Web Data Extraction

Bibliographic Data Extraction from the Web Using Fuzzy-Based Techniques

Tasleem Arif and Rashid Ali

1 Introduction

The exponential growth of the Web in addition to making the availability of diversified data ubiquitous [18], has made the job of information search very tough [26]. In order to find relevant information from the sea of data (the Web), the information scientists need to devise efficient techniques to extract the required data. These search techniques have to be supplemented with effective data processing techniques that can use the acquired data in an intelligent way that eventually can help address user queries satisfactorily. Thus, it can be assumed that the efficiency and efficacy of extraction, processing, and usage of data depends largely upon the methodology used by the end user/application to search for the desired results.

Though the use of search engines for finding relevant information on the Web is on the rise [22], technically, they are primarily limited to a simple relevance-ranking mechanism [11]. With the exponential growth of the Web and abundance of keywords repetition across documents the results returned by a search engine are vast and mostly beyond the limit of comprehension of a human being. Whereas, with the increasing role of named entities on the Web [6] it is expected that the search results should be *highly expressive*, e.g., if one searches for Prime Minister of a country, he should get a list of all the Prime Ministers of that country along with all other relevant details. Besides, the search results should present data

T. Arif (✉)
College of Computer Science and IT, Shaqra University,
Duwadmi, Riyadh Province, Kingdom of Saudi Arabia
e-mail: tarif@su.edu.sa

R. Ali
Department of Computer Engineering, Aligarh Muslim University,
Aligarh 202002, Uttar Pradesh, India
e-mail: rashidaliamu@rediffmail.com

© Springer Nature Singapore Pte Ltd. 2017
R. Ali and M. M. S. Beg (eds.), *Applications of Soft Computing for the Web*,
https://doi.org/10.1007/978-981-10-7098-3_7

extracted from various sources, such as Web pages, in an integrated fashion, and in a form that makes its analysis easy.

In the conventional search mechanism extracting contextual data appears quite difficult [8, 26]. This difficulty can be attributed to the absence of standardization of data, since the Web data can be structured, such as relations or relational tables; semi-structured, such as XML data; or, unstructured, such as spread sheets, text documents, CVs, Web pages, and presentations. The extent to which data can be classified as structured depends upon the extent to which it supports domain-specific queries. Web Data Extraction has been studied from the perspective of a broad range of application domains using different scientific tools [14]. The application of Web Data Extraction tools for Business and Competitive Intelligence, on the one hand, and for extraction of vast amounts of relationship data available on the social networks, on the other [14]. The latter has been largely responsible for attracting significant attention from the academics for analyzing these networks.

In addition to specific Web Data Extraction tools *wrappers* have traditionally been used to programmatically extract data of interest from a Web source. This data can be mapped to respective fields in a database table or an XML file [18]. Since the target Web source may contain a lot of information in a diversified fashion, the challenge that the wrappers face is to distinguish relevant data from the irrelevant.

The need for organized search has increased in the present context and the use of fuzzy-based techniques has enhanced the quality of search results [15]. Bibliographic data suffer from various problems like synonyms, missing data, typographic mistakes, different representations of similar information, etc. Exact techniques can suffer from low precision as well as low recall, thus decreasing the utility of the extracted data. In addition to these problems, in the recent years, the bibliographic data is being organized in domain specific databases like DBLP, PubMed, etc. Owing to these advances, organized fuzzy search can help to mitigate the problems faced by conventional search techniques in the bibliographic data extraction domain.

The rest of this chapter is organized in various sections. The second section discusses various publication features and their role in name disambiguation followed a discussion on issues related to extraction of publication data from Web sources, particularly Digital Citation Libraries. Section 4 deals with the proposed fuzzy mechanism and associated aspects. After that experimental results and discussions on the proposed data extraction and name similarity comparison are presented. The last section concludes the chapter.

2 Publication Features of Interest

The primary interest behind the extraction of publications data from online sources like digital libraries is proper profiling of authors. It is not unknown that there are a number of authors sharing same or similar names [4]. Thus proper attribution of

Table 1 Publication features used in author name disambiguation

Attribute	Description
$c_{i.}$ title	Title of the publication
$c_{i.}$ authors	Set of authors $\{a_1, a_2, ...\}$ of the publication
$c_{i.}$ e-mail IDs	Set of e-mail IDs $\{e_1, e_2, ...\}$ of authors
$c_{i.}$ affiliations	Set of affiliations $\{x_1, x_2, ...\}$ of authors
$c_{i.}$ conference/journal	Conference or journal name of the publication
$c_{i.}$ venue	Venue of publication
$c_{i.}$ yop	Year of publication
$c_{i.}$ references	References used by the publication
$c_{i.}$ contents	Contents of the publication

publications to their actual author is of utmost importance for efficient profiling and thus name disambiguation plays a pivotal role. The performance of any name disambiguation and profile integration mechanism depends upon the selection of publication features. It has been argued [23] that careful selection and use of publication features have a direct impact on the performance of a name disambiguation technique. In this section, we discuss various commonly used publication features and identify those which are important for name disambiguation. These features can be extracted from digital citation libraries, homepages of institutions and authors, PDF publications, etc. In the following section, we build on the discussions provided in [5]. Arif et al. [5] discussed various attributes along with their availability and usefulness. In this section, we elaborate on the commonly used attributes along with a discussion on additional promising features. Table 1 provides an overview of the conventional and additional features that may be used for the specified purpose.

2.1 Commonly Used Features

Various publication features, including author names, title, publication venue, date, and references, have been used for name disambiguation and author profiling either individually or in combination in the past [7]. Commonly used features include.

These features have been used either individually or in various combinations in the previous studies. Some features like co-authors have been used more commonly than others. A brief discussion of these features is provided in this section whereas, their availability and usefulness are discussed in detail in [5].

- **Publication Title**: Title of a publication is the primary identity of a citation-record. It defines the work carried out in the document in an abstract manner. The semantics of the title-string describe a number of things about the citation-record including the area of work, research interest of the author(s), interdisciplinary nature of the work, etc. Majority of the author name disambiguation approaches have leveraged the information contained in the title-string for accomplishing the job.

- **Author & Co-author**: It is one of the most widely used publication features for name disambiguation. Each publication has a principal author and zero or more co-authors. If two author names appear in a citation record together as authors, it is called as co-authorship relationship. It can be assumed that around 95% [1] of the publications have two or more authors and as such this information is very useful for name disambiguation.

- **Email IDs**: Email has been one of the most widely used communication media [3]. Even after the advent of Web 2.0, professional communication is largely dependent on the exchange of e-mails. In some ways, email ID of a person can be considered as a means of attributing some identity to a person. It is a well-known fact that email ID is a unique identification feature in the sense that no two persons can have exactly the same email ID, although one can have more than one email IDs [4].

- **Affiliations**: The name of the organization for which an author works is referred to as the *affiliation* of the author. This feature is made up of different components including name of the department, center, unit, group, laboratory, etc. besides the name of the organization and its address. This feature if used properly can help disambiguate ambiguous authors. Like email IDs, affiliations are also unique as no two affiliations can exactly be the same name. This feature is also useful in finding experts in a particular geographical location. The use of affiliations as disambiguation feature is based on the heuristic that "*no two different authors with same name have exactly the same affiliation.*"

- **Conference/Journal**: This feature identifies the *Journal* or the *Conference* in which the publication appeared. It is evident that majority of Journals publish articles in a particular research area which in itself is an authentication of the research interest of a particular researcher who publishes in that journal. Some conferences are a regular event and have certainly specified themes or focus areas, e.g., WWW Conference, SIGKDD Conference, etc. This again is an indication of the research interests of the author(s) of a publication in the proceedings of such Conferences. This publication feature is thus very beneficial in resolving conflicting research interest of ambiguous authors.

- **Venue**: The *venue* specifies the geographical location of publication of a publication. This feature is commonly used in case of Conferences, Seminars, Workshops, etc. It plays an important role in the analysis of research collaborations from a geographical perspective.

- **Year of Publication**: *Year of Publication* refers to the time at which the publication appeared. It is recorded as just the year in some journals and books, whereas in others it is recorded as the month and the year in which the publication first appeared. This timing information is more precise in case of conferences. From the perspective of social network analysis, it is important as it specifies the shifts in interest, affiliations, collaborations, etc. of authors and their research groups.

- **References**: *References* specify the material referred to by the author(s) of a publication. This appears as a separate section in all publications and contains the list of all the papers, books, etc. referred to in that publication. The use of this feature is based on the heuristic that "*no author refers to the work of another author with same name as of the author in question.*"
- **Contents**: It refers to the actual material contained in the publications. This can be used to derive a number of implicit conclusions about the area of work, research interest of the author(s), etc. This feature is rarely used in the author name disambiguation as it is very inefficient in obtaining the required information from the publication.

3   Metadata Extraction

Name disambiguation and profiling is not a straightforward task because its performance depends, *inter alia*, on quality of feature extraction methodology used. Extracting publication data from different sources is a resource intensive and tricky task. Institutional and personal homepages appear to be a promising source for extraction of publication data. Ciravegna et al. [12] have developed a system to mine affiliations of authors and their homepages by using DBLP and HomePageSearch.[1] The homepage of a target author can also be mined from the institutional website by making use of affiliation information. However, nonavailability of homepage information in digital citation libraries like DBLP and dependence on the existence of an external domain-specific resource [7] is one of the drawbacks of such an approach. There are other problems like absence of standardized format for publication of such data on an individual or institutional webpage [2].

In [2] publication data of four Indian Institutes of Technology (IIT) was extracted from their websites to explore the effectiveness and usefulness of publications data extracted from personal or institutional Webpage. Each IIT uses a different format for displaying their publications data on their respective websites. On the other hand, digital libraries like DBLP, CiteSeer, etc. store publications data in an organized format. Sometimes this data is made available for analysis purposes in the form of datasets, such as DBLP XML dataset.

The DBLP XML dataset is a huge file containing 1.2 Million publication records as of 2009 [19]. The size of full XML dump[2] in compressed format is *132* MB which upon extraction becomes more than *1* GB. Due to its huge size, it is not efficient, and sometimes not even possible, to extract and load the entire file into the memory and then parse XML bibliographic data for further processing. When we tried to use this file in our Java program on an Intel Core i5 1.80 GHz machine with

---

[1]http://hpsearch.uni-trier.de/.

[2]DBLP XML dataset: http://dblp.uni-trier.de/xml/dblp.xml.gz.

4 GB RAM, an exception *java.lang.OutOfMemoryError:Java heap space* occurred. This problem persisted even after making necessary changes in the size of the heap space.

In addition to the size of DBLP XML dataset, another important problem with using this dataset for extraction of bibliographic data is the age of this dataset. Since datasets are static snapshots of a database taken at a particular point in time, recent updates in the database are not visible in them. These problems could be mitigated by extracting bibliographic data dynamically. This approach overcomes both of these problems: there is apparently no serious problem with the size of the extracted bibliographic data for a given author name and it is also the most recent as it incorporates all the updates till that date.

4   Proposed Metadata Extraction Technique

In order to overcome the problems discussed above, we propose a Web data extraction tool to programmatically extract bibliographic data based on the name of the input author. The block diagram of the proposed data extraction approach is shown in Fig. 1. The proposed algorithm is outlined in Fig. 2. As pointed out in [14], many factors must be taken into account while designing a Web Data Extraction tool. These factors may or may not be dependent on the specific application domain, and as such, solutions which promise to be effective in one domain or context may not perform well in others.

The Web data extraction approach proposed here focuses on the HTML of the target author page retrieved from DBLP. The proposed approach uses the HTML tags that compose these Webpage along with their hierarchical organization to extract bibliographic information embedded in that page. In this case, specific HTML tags like *"This person"* have been used by DBLP to specify the target author, i.e., an author whom one is searching for. Such an approach which exploits the HTML structure and organization of Webpage is able to achieve a high level of accuracy and is also capable of scaling over a large collection of Webpage [14].

The proposed approach uses *jsoup* java HTML parser[3] that parses HTML to same Document Object Model (DOM) as modern browsers do.[4] Making use of the structure of the HTML tags, the algorithm extracts available publication features from publication records for each of the target author from DBLP publications database. After extraction, these features are stored in the local publications database if that publication record is not already available in the database.

The job of Step 3 and 4 of the algorithm outlined in Fig. 2 is the extraction of publications features for only those authors which have reasonable similarity with

---

[3]Jsoup Java Library: http://jsoup.org/download.
[4]http://jsoup.org/apidocs/.

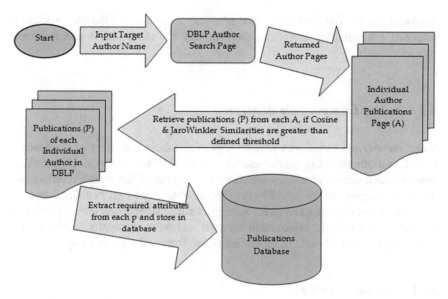

Fig. 1 Publications data extraction from DBLP

Input:	Target Author Name ($\mathcal{F}$)
Output:	Publication records of each author with the Target Author Name.

1. Retrieve from DBLP Author Search page A of all authors with given Target Author Name ($\mathcal{F}$)

2. *for* each retrieved author a (a belongs to A) in Step 1 *do*

3.   *if* CosineSimilarity(a, $\mathcal{F}$) >*Threshold*

4.   *if* JaroWinklerSimilarity(a, $\mathcal{F}$) >*Threshold*

5.   Retrieve from DBLP publications P of a

6.   *for* each p (p belongs to P)

7.   Extract authors, title, year_of_publication, journal/conference title

8.   Insert the extracted publication attributes in step 7 into database if that is not already available in database

9.   [End *for*]

10. [End *for*]

Fig. 2 Algorithm for extraction of publication features from DBLP

the target author. The commonly used similarity measures and the one used in the proposed bibliographic extraction methodology are discussed in the following section.

4.1 String Comparison Measures

String comparison techniques play an important role in a number of applications including disambiguating similar entities [13, 21, 25]. Bilenko et al. [9] provide a detailed classification of and discussion on the majority of the available techniques. These techniques can be broadly classified as character-based or token-based. In this section, we discuss some of the prominent string comparison techniques which are used, *inter alia*, in information retrieval and entity recognition. These techniques try to identify exact or approximate duplicates of a given strings.

4.1.1 Levenstein Distance

Levenstein distance is a basic edit-distance similarity measure which assigns a unit weight to all edit operations: less the number of edit operations, greater the similarity between two candidate strings. Thus, the difference between two candidate strings is the cost of edit operations required to convert one string to the other. The Levenstein distance $D(S, T, m, n)$ between first m letters in a string S and first n letters in another string T, as described in [9] is as follows (1):

$$D(S, T, m, n)$$
$$= \min \begin{cases} D(S, T, m-1, n-1), & \text{if } S_m = T_n, \text{and we copy } S_m \text{ to } T_n \\ D(S, T, m-1, n-1)+1, & \text{if we substitute } T_n \text{ for } S_m \\ D(S, T, m, n-1), & \text{if we insert the letter } T_n \\ D((S, T, m-1, n)), & \text{if we delete the letter } S_m \end{cases} \quad (1)$$

Needle–Wunsch distance [20], Smith–Waterman distance [24], etc. are some of the variations of the basic Levenstein distance that have also be used in some specific applications. In calculating Needle–Wunsch, distance, the different cost is associated with the three different edit operations viz., insertion, deletion, and substitution. In case of Smith–Waterman distance, cost of edit operations in the beginning of the string is the least and is most at the end of the string.

4.1.2 Jaro–Winkler Distance

Jaro–Winkler metric [27] is another important character-based string similarity measure. It is an extension of *Jaro* distance metric [16, 17], a popular record-linkage string comparison metric. The similarity score in this case

(Jaro distance) depends upon the ordering of characters along with the number of common characters between two candidate strings. The *Jaro* metric for two strings s and t is defined in Eq. (2).

$$\text{Jaro}(s,t) = \frac{1}{3} \cdot \left(\frac{|s'|}{s} + \frac{|t'|}{t} + \frac{|s'| - T_{s',t'}}{2|s'|} \right), \tag{2}$$

where s' is the number of characters in s that are common with characters in t in the same order as they appear in s, t' is the number of characters in t that are common with characters in s in the same order as they appear in t, and $T_{s',t'}$ is half the number of transpositions for s' and t' [9].

If the two candidate strings have common characters, in order, in the beginning, Winkler modification assigns higher weight to these matches, or vice versa. The Jaro–Winkler distance is defined as in Eq. (3).

$$\text{Jaro} - \text{Winkler}(s,t) = \text{Jaro}(s,t) + (lp(1 - \text{Jaro}(s,t))), \tag{3}$$

where l specifies the length of the longest common prefix of s and t, and p is a scaling factor (constant). In Winkler's implementation $l=4$ and $p=0.1$.

4.1.3 Jaccard Similarity

Jaccard Similarity is a simple but relatively effective token-based string comparison technique [9]. Jaccard-Similarity is the coefficient of the number of common tokens between any two strings S_1 and S_2 divided by a total number of tokens in both S_1 and S_2. The Jaccard-Similarity can be expressed as in Eq. (4).

$$\text{Jaccard} - \text{Similarity}(S_1, S_2) = \frac{(S_1 \cap S_2)}{(S_1 \cup S_2)} \tag{4}$$

4.1.4 Cosine Similarity

In certain situations, some features are more important than others and thus should be given more weightage in the comparison operation. *Term Frequency-Inverse Document Frequency (TF-IDF)* or *Cosine Similarity* provides more weightage to rare terms than to common terms in the two documents or strings. *TF-IDF*, which is a product of two frequencies, *Term Frequency (TF)* and *Inverse Document Frequency (IDF)*, has been widely used in information retrieval [9] is expressed as in Eq. (5).

$$\text{TF} - \text{IDF}(t_i, d_i) = \text{TF} \cdot \text{IDF}, \text{where} \tag{5}$$

$$\text{TF}(t_i, d_j) = \frac{n_i}{\sum_{j=1}^{N} n_j} \text{ and}$$

$$\text{IDF}(t_i, T) = \log \frac{|D|}{\text{DF}(t_i, d_i)},$$

where $\text{TF}(t_i, d_j)$ is equal to the number of times the term t_i appears in document d_j and $\text{DF}(t_i, d_i)$ is the number of times the term t_i appears in document d_i. D is the set of documents and T is the set having N terms.

5 Proposed Hybrid Similarity Measure

Character-based similarity metrics work efficiently for errors of data entry like typographical errors or abbreviations, whether inadvertent or deliberate (due to differences in the conventions being followed) but their efficiency decreases for larger strings [10]. Token-based similarity measures, however, work efficiently for larger strings treating them as bags of words. In token-based similarity measures, the order of the words in the string does not matter [10]. For use in author name disambiguation or publication metadata extraction, a combination of character-based and token-based similarity metrics can be explored. This work explores the use of character-based and token-based similarity measures in two stages for comparing various names and publication features.

In addition to its use in metadata extraction, we used the proposed methodology to calculate the similarity in three publication features viz. author names, affiliations, and publication venue titles for our name disambiguation techniques proposed earlier [4, 5]. To overcome the existing problems [5], we evaluated the use of a mix of similarity measures and different thresholds for different attributes. After testing a number of combinations and thresholds, we found that a combination of a token-based similarity measure and character-based similarity measure is more effective than using any one of these approaches in isolation. In our hybrid approach, we used Cosine-Similarity in the first stage and if the value for it was above a defined threshold, we used Jaro–Winkler Similarity in the second stage.

6 Experiments and Results

In this section, we present and discuss the experimental setup and results for name similarity computation and publication data extraction.

6.1   Name Similarity Computation

In this section, we show the results of experiments conducted for comparison of authors and affiliations. Table 2 presents a comparison of similarity values between two author name strings obtained by using commonly used string similarity measures. We compute the similarity values between two name strings using Jaccard, Cosine, and Jaro–Winkler string similarity measures.

Analysis of the values presented in Table 2 clearly reveals that no single similarity/distance function can decide whether two name strings are similar or different. It is evident from the values listed in this table that Jaccard-Similarity measure fails in a number of cases, where exactly similar values are obtained for two exactly similar strings and two different strings. If we take into account Cosine and JaroWinkler similarity values, there are some cases where the name refers to the same person but it has low JaroWinkler similarity and high cosine similarity whereas, in some others, we have low Cosine similarity and high Jaro–Winkler similarity. Thus, it can be concluded that neither Cosine similarity nor Jaro–Winkler similarity is capable of finding a matching or different author name individually. In order to overcome these problems, we use hybrid similarity measure to compare author names and affiliations, where we use cosine similarity in the first stage and Jaro–Winkler in the second stage only when cosine similarity is above a threshold.

Table 2 Similarity values of different string similarity measures

String 1	String 2	Similarity score		
		Jaccard	Cosine	Jaro–Winkler
M. Asger	Asger M.	1.00000	1.00000	0.80952
Mohammed Asger	M. Asger	0.33333	0.50000	0.54762
M. M. Sufyan Beg	M. Asger	0.25000	0.57735	0.53571
Bing Liu	B. Liu	0.33333	0.50000	0.82750
Bing Liu	Lin Liu	0.33333	0.50000	0.86905
Rakesh K. Kumar	Rakesh Kumar	0.66667	0.81650	0.97959
Rakesh K. Kumar	R. K. Kumar	0.33333	0.70711	0.47222
Rakesh K. Kumar	A. Kumar	0.00000	0.00000	0.55714
Rakesh S. Kumar	Rakesh Kumar Singh	0.50000	0.66667	0.87684
Yenji Tang	Jie Tang	0.33333	0.50000	0.80833
J A Walsh	Ajay Gupta	0.00000	0.00000	0.53148
A G Sharpe	Ajay Gupta	0.00000	0.00000	0.60000
Rashid Ali	Rashid Al-Ali	0.33333	0.50000	0.97633
Rashid Ali	Rashid J. Al-Ali	0.25000	0.40825	0.92969
Wajid Ali Khan	Rashid Ali	0.25000	0.40825	0.73333
Wahid Ali Khan	Ali Shaikhali	0.25000	0.40825	0.79451
Jin Zhang	Qian Zhang	0.33333	0.50000	0.81296
Jin Zhang	Qing-Yu Zhang	0.25000	0.40825	0.70976

Table 3 Prediction results of the proposed approach for author name similarity computation

String 1	String 2	Prediction results	
		Predicted	Actual
M. Asger	Asger M.	Same	Same
Mohammed Asger	M. Asger	Same	Same
M. M. Sufyan Beg	M. Asger	Different	Different
Bing Liu	B. Liu	Same	Same
Bing Liu	Lin Liu	Same	Different
Rakesh K. Kumar	Rakesh Kumar	Same	Same
Rakesh K. Kumar	R. K. Kumar	Same	Same
Rakesh K. Kumar	A. Kumar	Different	Different
Rakesh K. Kumar	Rakesh Kumar Singh	Same	Same
Yenji Tang	Jie Tang	Similar	Different
J A Walsh	Ajay Gupta	Different	Different
A G Sharpe	Ajay Gupta	Different	Different
Rashid Ali	Rashid Al-Ali	Same	Different
Rashid Ali	Rashid J. Al-Ali	Different	Different
Wajid Ali Khan	Rashid Ali	Different	Different
Wahid Ali Khan	Ali Shaikhali	Different	Different
Jin Zhang	Qian Zhang	Same	Different
Jin Zhang	Qing-Yu Zhang	Different	Different

The proposed string similarity computation uses the concept of approximate (fuzzy) matching in combination with token-based string comparison to derive the final score.

The results of our hybrid name comparison similarity methodology are shown in Table 3. From the analysis of the prediction results, it can be observed that the proposed hybrid (fuzzy) string comparison methodology is capable of achieving a high degree of prediction accuracy. The proposed string comparison has been used successfully for comparing affiliations of authors used in scientific publications. The results and the methodology for comparing affiliation(s) of authors have been discussed in [5] in detail.

6.2 Publications Metadata Extraction

Figure 3 shows a snapshot of the intermediate output produced by the proposed approach wherein all the authors with the target author name are retrieved from DBLP database. The proof of the authenticity of the output shown in this snapshot can be verified from Fig. 4 as all those authors which have been shown as a search result in Fig. 4 are also listed in intermediate results shown in Fig. 3.

Fig. 3 Parsed DBLP author links for '*Rashid Ali*'

Figure 4 shows the search results for the author name '*Rashid Ali*' on DBLP. From this figure, it is clear that "*Rashid Ali*" appears in one format or the other, either wholly or partially in nine different author names.

Figure 5 shows a snapshot of the intermediate parsing results when individual publication features of a publication record are extracted. This figure lists various authors of the current publication record, its title, year of publication, and conference/journal in which this publication appeared. It can be observed from this figure that our approach works quite efficiently for extraction of listed publication features from publications page of the author under consideration from DBLP.

Figure 6 shows a snapshot of DBLP author bibliographic page. This snapshot is for the author name "*Rashid Ali*". From the analysis of Fig. 6, it can be observed that the bibliographic results obtained for author name "Rashid Ali" from DBLP suffer from mixed citation problem. The citation-record listed as a first result for the year 2009 with title "*A wavelet method for solving singular integral equation of MHD*" does not belong to that Rashid Ali to which other publications belong to. This is a case of mixed-citation. Similarly, there are cases of split-citation also. We observed that these problems exist in case of a number of other authors also. This clearly indicates that the mixed citation problem is prevalent in digital libraries like DBLP.

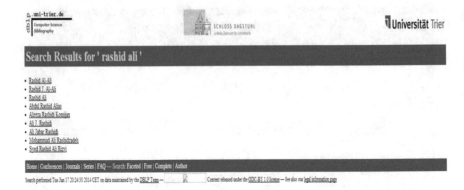

Fig. 4 DBLP search results for '*Rashid Ali*'

Fig. 5 Feature extraction from individual DBLP publication records

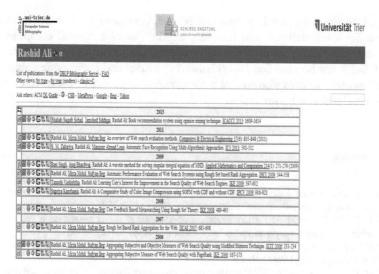

Fig. 6 DBLP author (for *Rashid Ali*) bibliographic webpage

7 Conclusions and Future Directions

The Internet is a vast repository of information. This information may be structured or unstructured. The relevance and usefulness of the extracted information depend largely on the methodology used. In the majority of the cases, it has been observed that the general purpose search engines overload the user with irrelevant responses. Thus the extraction of desired information largely depends upon the expertise of the person and the way he handles the generated responses. But generalized search may not be useful in a number of circumstances. In such cases, one needs specialized tools to extract the desired information effectively and efficiently.

The presence of named entities on the Web is increasing at an enormous pace and our dependence on computer-based information systems that deal with named entities has also increased. Thus, we need a strong mechanism to determine the difference or similarity between these entities. There are a whole lot of both exact and fuzzy string matching techniques. It has been observed that the efficiency of these measures vary from situation to situation and there is no such mechanism that fits all. However, if we combine two or more such measures the efficiency of the matching task can be improved to a great extent. This chapter introduced one such combination and presented the efficiency of the proposed fuzzy string matching technique. Though we have used static thresholds for comparisons, variable thresholds can also be explored for better results.

This chapter also presented a technique for extraction of live publications data from digital citation libraries using the proposed fuzzy string matching technique. It was observed that such a technique which extracts live publications data has the advantage of performing better over static datasets on two fronts. First, the extracted data reflects the most recent changes and second, the size of the data extracted is not beyond the computational and handling capabilities of a normal computer. The proposed data extraction methodology extracted data from only one source, i.e., DBLP but it can be modified and enhanced to extract and incorporate publications data from other sources also so that a comprehensive picture of the research profile of an individual can be presented.

References

1. Ahn J, Oh D, Lee J (2014) The scientific impact and partner selection in collaborative research at Korean universities. Scientometrics 100(1):173–188
2. Arif T, Ali R, Asger M (2012) Scientific co-authorship social networks: a case study of computer science scenario in India. Inter J Comput Appl 52(12):38–45
3. Arif T, Ali R, Asger M (2014) Social network extraction: a review of automatic techniques. Inter J Comput Appl 95(1):16–23
4. Arif T, Ali R, Asger M (2014) Author name disambiguation using vector space model and hybrid similarity measures. In: Proceedings of 7th international conference on contemporary computing, IEEE Press, Noida, India, pp 135–140
5. Arif T, Ali R, Asger M (2015) A multistage hierarchical method for author name disambiguation. Inter J Inf Process 92(3):92–105
6. Artiles J, Amigo E, Gonzalo J (2009) The role of named entities in web people search. In: Proceedings of the 2009 conference on empirical methods in natural language processing, Singapore, 6–7 August 2009, pp 534–542
7. Aswani N, Bontcheva K, Cunningham H (2006) Mining information for instance unification. In: Proceedings of 5th international semantic web conference, Athens, GA, USA, pp 329–342
8. Barrière C (2016) Natural language understanding in a semantic web context. In: Searching for named entities, pp 23–38
9. Bilenko M, Mooney R, Cohen W, Ravikumar P, Fienberg S (2003) Adaptive name matching in information integration. IEEE Intell Syst 18(5):16–23
10. Bilenko M, Mooney RJ (2003) Adaptive duplicate detection using learnable string similarity measures. In: Proceedings of the 9th ACM SIGKDD international conference on knowledge discovery and data mining, Washington, USA, pp 39–48
11. Cafarella MJ (2009) Extracting and managing structured web data. Ph.D. Thesis, Department of Computer Science & Engineering, University of Washington, USA. http://turing.cs.washington.edu/papers/cafarella_thesis.pdf
12. Ciravegna F, Chapman S, Dingli A, Wilks Y (2004) Learning to harvest information for the semantic web. In: Proceedings of the 1st European semantic web symposium, Heraklion, Greece, pp 312–326
13. Das J, Choong PL (2007) Resolving partial name mentions using string metrics. Defence Science and Technology Organisation, PO Box 1500, Edinburgh, South Australia, 5111-Australia. Available at http://www.dsto.defence.gov.au/corporate/reports/DSTO-RR-0318.pdf
14. Ferrara E, De-Meo P, Fiumara G, Baumgartner R (2014) Web data extraction, applications and techniques: a survey. Knowl-Based Syst 70:301–323
15. Hoeber O, Yang XD (2006) Visually exploring concept-based fuzzy clusters in web search results. In: Proceedings of the Atlantic web intelligence conference, 2006
16. Jaro MA (1989) Advances in record-linkage methodology as applied to matching the 1985 census of Tampa, Florida. J Am Stat Assoc 84:414–420
17. Jaro MA (1995) Probabilistic linkage of large public health data files. Stat Med 14:491–498

18. Laender AHF, Ribeiro-Neto BA, Da-Silva AS, Teixeira JS (2002) A brief survey of web data extraction tools. ACM SIGMOD Rec 31(2):84–93
19. Ley M (2009) DBLP-some lessons learned. The Proc VLDB Endow 2(2):1493–1500
20. Needleman SB, Wunsch CD (1970) A general method applicable to the search for similarities in the amino acid sequences of two proteins. J Mol Biol 48:443–453
21. Rees T (2014) Taxamatch, an algorithm for near ('Fuzzy') matching of scientific names in taxonomic databases. PLoS One 9(9):e107510. https://doi.org/10.1371/journal.pone.0107510
22. Rosenfeld MJ, Thomas RJ (2012) Searching for a mate: the rise of the internet as a social intermediary. Am Sociol Rev 77(4):523–547
23. Smalheiser NR, Torvik VI (2009) Author name disambiguation. Ann Rev Inf Sci Technol 43(1):1–43
24. Smith T, Waterman M (1981) Identification of common molecular subsequences. The J Mol Biol 174(1):195–197
25. Sun Y, Ma L, Wang S (2015) A comparative evaluation of string similarity metrics for ontology alignment. J Inf Comput Sci 12(3):957–964
26. Trotman A, Zhang J (2013) Future web growth and its consequences for web search architectures. 2013. arXiv:1307.1179v1
27. Winkler WE (1999) The state of record linkage and current research problems. Statistics of Income Division, Internal Revenue Service Publication R99/04. Available at http://www.census.gov/srd/www/byname.html

Part IV
Soft Computing Based Question Answering Systems

Crop Selection Using Fuzzy Logic-Based Expert System

Aveksha Kapoor and Anil Kumar Verma

1 Introduction

Despite being one of the foremost professions since time immemorial, agriculture [1] has still been largely untouched by the modern information technology. There exists a wide communication gap among a large proportion of farmers regarding the latest agricultural research on new varieties of seeds and technologies to sow them. Moreover, farming requires a lot of decision-making right from planting, growing, to the harvesting of crops. During the entire course of a crop, a farmer faces a lot of challenges like crop selection, and once the selected crop is planted, preventive and curative measures against pests, diseases, natural forces (like frost, extreme hot, etc.) are required. To top that, many-a-times farmers only have partial knowledge about the techniques required to face all these challenges which cause wrong decision-making and thus incorrect outcomes resulting in drastic losses in the form of crop yield [2].

If efforts are made to automate the decision-making process at every stage of a crop, then the farmer will only need to concentrate on the what to do rather than how and what to do at each phase of a crop. This work is an attempt to automate the first phase, i.e. crop selection for a farmer through the use of a fuzzy logic-based expert system. Based on certain parameters according to the climatic and soil conditions, a farmer can provide some input data and a crop name is obtained as the output. Some additional tips regarding sowing techniques, machinery and reliable government approved vendors can also augment the crop name in the output.

A. Kapoor (✉) · A. K. Verma
Computer Science and Engineering Department, Thapar University, Patiala, India
e-mail: veke095@gmail.com

A. K. Verma
e-mail: akverma@thapar.edu

© Springer Nature Singapore Pte Ltd. 2017
R. Ali and M. M. S. Beg (eds.), *Applications of Soft Computing for the Web*,
https://doi.org/10.1007/978-981-10-7098-3_8

1.1 Fuzzy Logic

Fuzzy logic stands for human-like logic or reasoning [3]. The humans rarely talk or devise their day-to-day rules in terms of Boolean logic. Rather, our rules are of more approximate and probabilistic nature with no sharp distinction between the boundaries of different hypothesis and results. In simple terms, two or more hypothesis can be partly true to give partly true results. Furthermore, our rules are not limited to true or false; we have a set of ranges.

For instance, if we want to access whether the mileage of a certain car is good (fuzzy set) as in Fig. 1. A traditional machine will answer in true or false. On the contrary, a human will answer it as excellent, most likely good, sometimes good, average, mostly bad (linguistic variables) and we also want our machines to understand and give results according to our fuzzy logic like interpretations. It is because machines have been made for solving our purposes rather than causing an overhead for understanding their Boolean interpretations and then converting it into our own human-like logic.

Moreover, it is the fuzziness which creates compactness of values, rules and belief systems we as humans have set for ourselves. If we had mathematical input and output mapping for all our rules, there would rather be a lot of unnecessary complexities in our brains. Rather, the concept of fuzzy logic or defining everything in vague terms is what makes our functioning a lot more simpler. The strength of fuzzy sets lies in the allowance of an overlap amongst the different linguistic variables [4]. Hence, in the car example of Fig. 1, an overlap between average and sometimes good can occur implying that the car mileage is partly a member of 'average' as well as 'sometimes good' ranges.

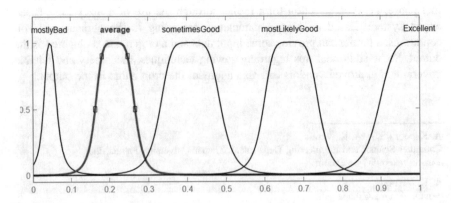

Fig. 1 Fuzzy set 'mileage'

1.1.1 Significance of Fuzzy Logic in Crop Selection

In this paper, the authors have implemented fuzzy rules-based crop selection system. The decision to use fuzzy logic was taken on the basis of many factors. The rules have rendered a compact and more human-like expert structure without losing their accuracy which will be discussed in the results Sect. 5. An effort to decode the crop name as a function of many climatic and soil properties has been made. Using mathematical ranges say, for the amount of rainfall, soil pH would have made rules non-intuitive to a human which would have caused unnecessary overhead while customizing or adding rules. Computing with words [3] using fuzzy logic will help farmers, agriculturists to use and customize such systems without a need for extra training to understand the computer and mathematics related technicalities of the system. Taking for instance, in input data, if we had taken mathematical ranges for rainfall, each time to narrow down the range or widen it would have required lots of redundant efforts. On the other hand, having fuzzy rules helps as in all forms of widening and narrowing down in the ranges do not affect the wording of the rules at all (only internal re-factoring of ranges is required).

2 Motivation

Every year, many kinds of crops get damaged even before the harvest can be reaped. The major reason is the inability of a farmer to predict what kind of crops to be grown according to the changing topography, nutrient content, pH value, salinity, water content, climate conditions and consequently, leading to wrong predictions regarding the type of crops to plant this season, the amount of water required, upcoming weather conditions and the specific techniques to be used for a certain crop. The knowledge regarding all these parameters exists but is many-a-time limited to national organizations like Tropical Agricultural Association (TAA), Fertilizer Association of India or to some very experienced farmers. Hence, a redistribution and dissemination of this agricultural knowledge are required. Moreover, even these national organizations mostly specialize only in one narrow domain. On the contrary, a farmer needs knowledge in all domains to have productive yield.

One way is to educate all the farmers over the years and if for all the new techniques arise, incremental session of knowledge building and exchange should occur. Another better and quicker method is to develop a common fuzzy expert system which will answer the type of crop, the amount of water required, some precautions according to weather predictions and some basic tips in response to various parameters (like pH, salinity, water content, type of clouds—nimbus, stratus, etc. at that time) input by the farmer in a fuzzy human-like form.

The system proposed has a wide range of scope and applications which include:

- Knowledge sharing among farmers all over India through knowledge population in the common expert system proposed.
- Quicker Farming research dissemination by the researchers to the farmers through enhancement of rule-base of Crop Selection Expert System.
- Overall increase in crop quality, farmer income, living standard, better export quality and increase in national GDP.
- No extra burden on farmer to go through intensive guidance programmes, as they will now be equipped with a simple question–answering system.
- Integration of knowledge of all kinds of units be it seeds associations, fertilizer associations, irrigation techniques associations into one rule base.

In addition, as a part of Digital India plan, the Government of India [1] is taking steps to provide guidelines for farming according to specific region. There are other state-level agricultural universities which provide a lot of data relevant to farming. The opportunity lies in the selection of appropriate crop out of this humongous data. Therefore, expert systems which can capture this data and provide it to farmer in the form of a rule-based decision system for crop selection, pest management and harvesting tips, etc. are inevitable. The authors work is an effort to initiate this task with a model crop selection expert system.

3 Literature Review

A lot of efforts in the form of various expert systems, research on agricultural practices, seed varieties and impetus by the Government of India are assisting the farmers. The 'Ministry of Agriculture and Farmers Welfare' [1] has made commendable efforts of serving the entire agriculture-related data over the web by enabling farmers to choose their region and thus providing monthly guidelines for their crops in their native language. Moreover, the website has served as an important source of knowledge acquisition for the crop-related data in this work. There still exists an opportunity to have web-based expert systems for crop selection, pest management, etc. on this website.

Ladha et al. [2] have discussed factors causing a long-term decline in the yield of major crops like rice and wheat in Asia (with India having a special mention) due to usage of incorrect amount of fertilizers and other climate change factors. The decline in K (Potassium) in soil has been reported as the major reason behind decrease in rice yields specifically. Hence, farmers need to be well acquainted with the changing climate and thus updates in the agricultural techniques. The proposed system takes into account the soil nutrient content be it naturally or from fertilizers, climatic conditions and irrigation facilities to suggest a suitable crop. Whatever changes in farming techniques are required, they can be incorporated into the system proposed in this paper; thus enabling farmers to receive maximum yield and profits from the knowledge updates in this expert system; despite the transient nature of topography and biosphere.

Montalvo et al. [5] have elucidated a novel idea of image-processing-based expert system to detect weeds in a maize field using image-segmentation techniques. The system achieved a maximum success of 93.4% and a minimum success rate of 85.3%. Ponnusamy et al. [6] analyse the effectiveness of a cattle and buffalo expert system in skill acquisition among the dairy farmers. The results suggest that greater willpower and motivation among the commercially inclined big dairy farmers than the marginal farmers play an important role in the effectiveness of the expert system. This work serves as an important guideline for all the concerned authorities to motivate and educate the farmers by advertising, direct training regarding the use of tried and tested expert systems. Building upon the ideas presented in these works, the authors of the current paper have also tried to make things easier for farmers by building most of the tedious decision-making into a software system (as farmers have a lot of others purposes to look into through the entire life-span of nurturing a crop).

Hasan et al. [7] devised a web-based expert system for timely diagnosis of sugarcane-related diseases. This is a great step towards assisting farmer once the suitable crop has been selected, planted and is then undergoing maturity in the field. The authors claim that their system is very customizable for other crops too. Our work complements their future work by proposing a system to first plan the crops to be grown this season and then have a diagnostic system once the crop has been planted. Adekanmbi et al. [8] look at optimal crop planning from an economic point of view, considering factors like land, fixed and variable cost. Thus, an expert system to determine the crop-mix was developed using some mathematical optimization techniques. The system was developed for local farmers in South Africa and changing climatic factors were not a part of crop planning. Our system considers the perspective of changing biosphere and can enhance the former system by taking into account biosphere changes along with efficient land allocation for maximum crop yield.

Kawtrakul et al. [9] modeled a personalized planning system for Thailand farmers to determine the variety of rice they should grow based on factors like climate and diseases in the area. This paper provides sufficient opportunity for the authors to develop a generalized crop selection system with India in focus.

Zadeh [3] is the father of fuzzy logic. His work provides an in-depth understanding of why and how to use fuzzy logic in different situations. MATLAB Fuzzy Logic Toolbox [10] is an indispensable tool to write fuzzy rules and analyse them using a user-friendly GUI (Graphic User Interface). Negnevitsky's [4] book on artificial intelligence is a practical approach to the understanding artificial intelligence and expert systems using many real-life experiments which have helped authors get a better grip over fuzzy logics and essence of expert systems in general. Furthermore, one of the challenges faced was knowledge acquisition for the crop selection expert system is that very less authentic and reliable research material is available regarding the climate and soil suitability for different crops. A lot of data is not recorded in written form and was extracted out by interviewing various expert farmers. Hence, the proposed system is also an attempt to record all the scattered and verbally available knowledge. Dhaliwal et al. [11], a 2014 publication of

Punjab Agricultural University provides an interesting insight into the different factors, techniques taken into account throughout the entire course of a crop. This book provides useful tips to the farmers for growing different types of crops in Punjab, India. It has been one of the motivations for the authors to collect and structure this data in the form of some expert system where farmer does not have to read and understand so many theoretical nuances of agriculture. Also, Joshi [12], Datta [13], Venugopalan et al. [14], 'Ministry of Agriculture and Farmers Welfare' [1] provide data on maize, rice, bt cotton and some other crops, respectively.

Balezentiene et al. [15] is an interesting read on farming yield regarding biomass fuel generation. The paper discussed about a system to find the crop giving maximum biomass yield which is a renewable energy source. Jawad et al. [16] have discussed the yield of a given crop as a function of humidity, temperature and rainfall in an area. This will help farmers predict which crop to grow based on its yield. On the other hand, our system needs no crop input; it takes into account many other factors like soil pH, nutrient content, type of soil besides humidity, temperature and rainfall.

4 Proposed Work Methodology

4.1 Tools and Softwares Used

MATLAB Fuzzy Toolbox has been used by the authors for designing the expert system. The fuzzy rules were generated from multitude of research papers in this field, verbal question–answering with expert farmers and data gathered from the publications of various agricultural universities throughout India as discussed in the Literature Review Sect. 3.

4.2 Choosing the Input Parameters

A lot of analysis was done to choose the final list of input parameters. After reading multitude of literature on various crops in India, some common parameters which could fully determine the name of crop to be grown were found. The authors have attempted to work on the analogy that as a certain hash requires a unique set of values, similarly, there exists a mapping between a crop and the unique set of favourable natural factors. These favourable natural factors serve as our input parameters. Here, only the natural factors that best favour a crop have been considered because after all crop selection should be such that it cause the highest yield from amongst all the other alternatives. The initial set of input parameters proposed were:

1. Rainfall
2. Temperature
3. Humidity
4. Sunshine
5. Cloud type
6. Nitrogen (cumulative nitrogen amount available in soil and fertilizers)
7. Calcium (cumulative calcium amount available in soil and fertilizers)
8. Phosphorous (cumulative phosphorous amount available in soil and fertilizers)
9. Potassium (cumulative potassium amount available in soil and fertilizers)
10. Humus
11. Iron (cumulative iron amount available in soil and fertilizers)
12. Zinc (cumulative zinc amount available in soil and fertilizers)
13. Soil texture
14. Soil type
15. Topography type (plains, mountains, valleys, etc.)
16. Soil pH
17. Soil depth (depth of top fertile layer of soil)
18. Moisture holding capacity of soil
19. Irrigation possible (amount of irrigation the farmer can provide)
20. Amount of labour available
21. Cloud type
22. Soil salinity
23. Region of India (north-east, north, south, etc.)

Out of these cloud type, soil salinity and region of India were found to be redundant parameters. Soil salinity is already covered in the type of region like a salt desert region and they are generally not favourable for crops. For instance, crops like date palm can tolerate salinity but they grow in deserts, so other parameters suffice in making decisions. Region of India is a wrong choice as

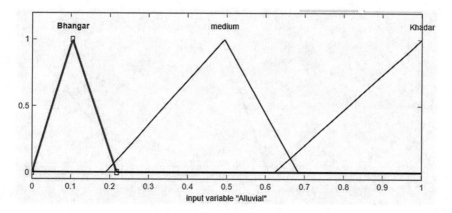

Fig. 2 Alluvial soil

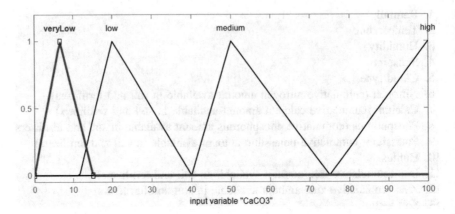

Fig. 3 CaCO₃ content

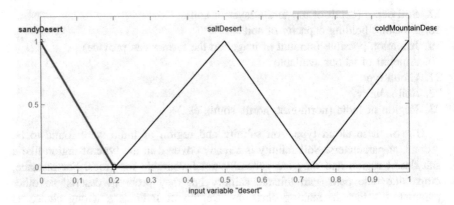

Fig. 4 Desert region

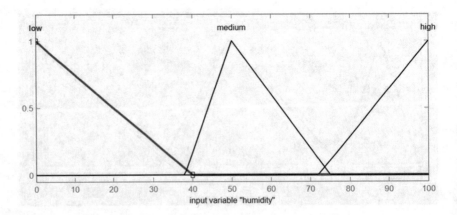

Fig. 5 Humidity content as %

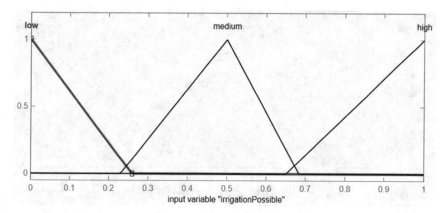

Fig. 6 Amount of irrigation available in terms of probability

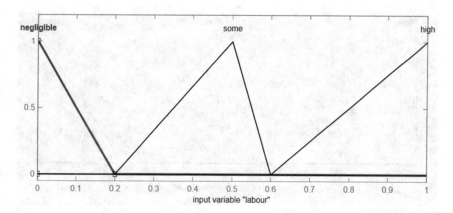

Fig. 7 Amount of labour required

climate of some places is constantly changing like plains converting to deserts and vice versa. So topography serves as a better choice for input parameter. Cloud type is determined by temperature, humidity and other weather conditions already covered in other parameters. Figures 2, 3, 4, 5, 6, 7, 8, 9, 10, 11, 12, 13, 14 and 15 show some of these input variables and their corresponding membership functions (triangular type). The mfs (membership functions) for other nutrients have been modeled similar to Fig. 3.

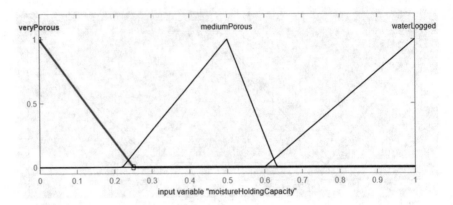

Fig. 8 Moisture holding capacity of soil

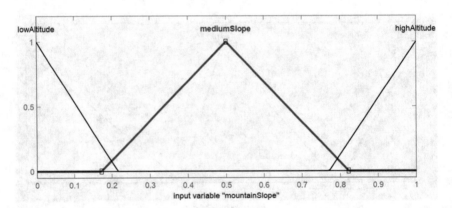

Fig. 9 Altitude of mountain slope

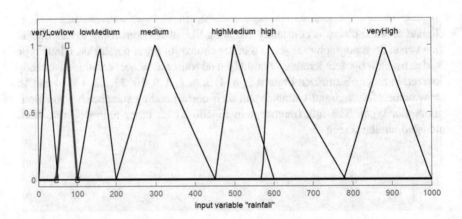

Fig. 10 Amount of rainfall (in cm)

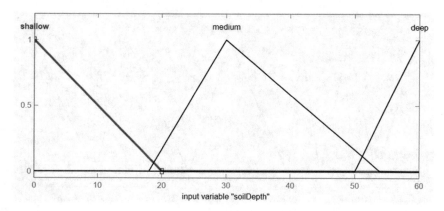

Fig. 11 Depth of top soil

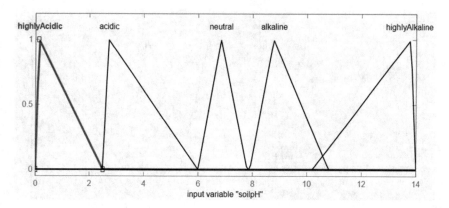

Fig. 12 Soil pH value (1–7)

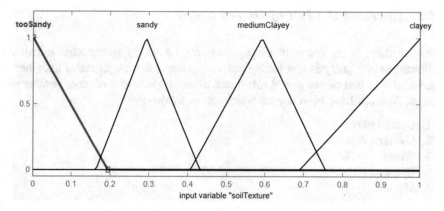

Fig. 13 Soil texture

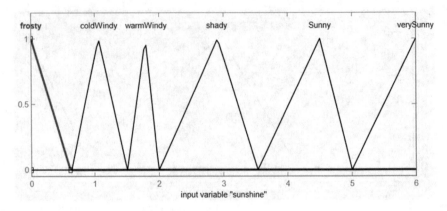

Fig. 14 Amount of sunshine required

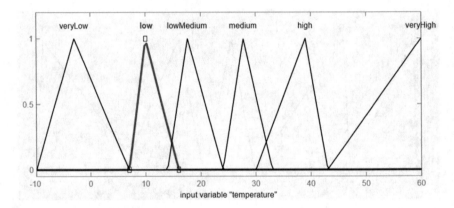

Fig. 15 Temperature conditions (in °C)

4.3 Choosing the List of Output Crops

Though there is no dearth of crops which can be grown specifically in Indian climate; yet for analysis and testing purposes, only a few major crops have been included as a part of our model rule-based expert system. Out of large number of crops, 20 crops have been chosen which are as follows:

1. Upland rice
2. Lowland rice
3. Wheat
4. Gram
5. Maize
6. Jowar

7. Bajra
8. Cotton
9. Jute
10. Groundnuts
11. Sesame
12. Sunflower
13. Mustard
14. Linseed
15. Castor seed
16. Rubber
17. Sugarcane
18. Tea
19. Coffee
20. Ragi

4.4   Formulating the Rules

Figure 16 shows only some of the total 27 rules devised for 20 crops. For some crops, there are many alternative input conditions. For example, considering 'jowar', the rules are:

If (rainfall is low) and (moisture Holding Capacity is medium Porous) and (temperature is medium) and (soil is black Soil) and (topography is plateau) and (labour is negligible) then (crop is jowar) If (rainfall is low) and (moisture Holding Capacity is very Porous) and (temperature is medium) and (soil is desert Soil) and (topography is plains) and (labour is negligible) then (crop is jowar)

11. If (rainfall is low) and (sunshine is verySunny) and (moistureHoldingCapacity is mediumPorous) and (temperature is medium) and (C;
12. If (rainfall is lowMedium) and (Alluvial is medium) and (moistureHoldingCapacity is mediumPorous) and (temperature is medium) and (
13. If (rainfall is low) and (sunshine is not frosty) and (moistureHoldingCapacity is mediumPorous) and (temperature is lowMedium) and
14. If (rainfall is veryLow) and (sunshine is not frosty) and (temperature is lowMedium) and (plains is plains) and (loamy is loamySoil) th
15. If (rainfall is veryLow) and (sunshine is not frosty) and (temperature is lowMedium) and (plains is plains) and (BlackSoil is BlackSoil)
16. If (rainfall is lowMedium) and (temperature is lowMedium) and (plateau is plateau) then (crop is sunflower) (1)
17. If (rainfall is veryLow) and (Alluvial is medium) and (moistureHoldingCapacity is mediumPorous) and (temperature is low) and (plains
18. If (rainfall is low) and (Alluvial is medium) and (temperature is lowMedium) and (plains is plains) then (crop is linseed) (1)
19. If (rainfall is low) and (temperature is lowMedium) and (BlackSoil is BlackSoil) and (plateau is plateau) then (crop is linseed) (1)
20. If (rainfall is low) and (sunshine is not frosty) and (temperature is lowMedium) and (RedSoil is RedSoil) and (desertSoil is desertSoil)
21. If (rainfall is medium) and (soilDepth is deep) and (sunshine is verySunny) and (moistureHoldingCapacity is mediumPorous) and (tem
22. If (rainfall is medium) and (Alluvial is Khadar) and (soilpH is neutral) and (sunshine is not frosty) and (moistureHoldingCapacity is me
23. If (rainfall is lowMedium) and (soilDepth is deep) and (sunshine is shady) and (moistureHoldingCapacity is mediumPorous) and (temp
24. If (rainfall is lowMedium) and (sunshine is shady) and (moistureHoldingCapacity is mediumPorous) and (temperature is medium) and
25. If (rainfall is medium) and (sunshine is shady) and (moistureHoldingCapacity is mediumPorous) and (temperature is medium) and (Ca
26. If (rainfall is medium) and (sunshine is shady) and (moistureHoldingCapacity is mediumPorous) and (temperature is medium) and (Ca

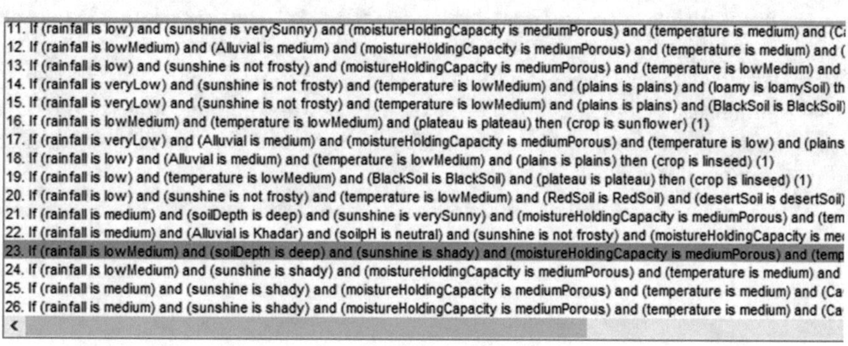

Fig. 16 A snapshot of fuzzy rules

The MATLAB Fuzzy ToolBox does not allow 'or' operation in a rule. Hence, it needs to be written separately. Hence, existence of more than one rule for some crops in case of some 'or' condition accounts for the number of rules exceeding the number of crops. Also, some rules have been shown in tweaked format so as to include the output crops in the view and get a better understanding of the rules. Rules about some selected crops have been discussed as follows:

1. If (rainfall is low) and (soil is red or black or desert Soil or mountain Soil) and (soil pH is highly acidic) and (soil Depth is shallow) and (sunshine is very sunny) and (moisture Holding Capacity is very porous) and (temperature is low) and (nitrogen availability is high) and (phosphorous availability is medium) and (potash availability is medium) and (humidity is low) and (topography is sandy desert or plateau or east Coast or valley or medium mountain slope) and (irrigation Possible is low) and (iron is low) then crop is **upland rice**.

2. If (rainfall is very low) and (soil is alluvial) and (sunshine is very sunny) and (moisture Holding Capacity is waterlogged) and (temperature is low medium) and (nitrogen availability is high) and (phosphorous availability is medium) and (irrigation Possible is high) and (labour is high) then crop is **lowland rice**.

3. If (rainfall is low) and (soil is medium bhangar alluival) and (soil pH is neutral) and (moisture Holding Capacity is medium porous) and (temperature is low) and (topography is plains) and (nitrogen availability is high) and (phosphorous availability is medium) and (potash availability is medium) and (irrigation Possible is low) and (labour is medium) and (loam quality in soil) and (soil Texture is clayey) then crop is **wheat**.

4. If (rainfall is medium) and (soil Depth is deep) and (sunshine is very sunny) and (moisture Holding Capacity is medium porous) and (temperature is medium) and (humidity is high) and (soil is laterite) and (mountain Slope is medium) and (labour is high) and (loam quality in soil) then crop is **rubber**.

5. If (rainfall is medium) and (sunshine is shady) and (moisture Holding Capacity is medium porous) and (temperature is medium) and (calcium carbonate availability is high) and (soil is red) and (mountain Slope is high altitude) and (labour is high) and (loam quality in soil) and (soil Texture is medium clayey) and (iron availability is high) and (humus is high) then crop is **coffee**.

5   Results and Discussion

The proposed system has been tested by picking an arbitrary region and recording its current climatic conditions and soil properties. Then we apply the fuzzy rules and find the best match to get the output crop. The output is compared with the actual crop grown at that time with the given climate and soil properties for the region in question. Suppose there is a following type of area:

A humid area with heavy rainfall of around 1000 mm, very less sunshine with shady trees and sandy loam deep soil with average porosity. It has a mild climate with the temperature staying from 10 to 20 °C and lots of NPK fertilizers are available to the farmers. There is a scope for large labour employment in this mountainous region at an altitude of around 1500 m. What crop should I grow?

So formulating the rule (considering membership functions' figures in Sect. 4.2):

If (rainfall is low Medium) and (soil Depth is deep) and (sunshine is shady) and (moisture Holding Capacity is not water Logged) and (temperature is low Medium) and (Phosphorous availability is high) and (Potash availability is high) and (humidity is high) and (labour is high) and (soil Texture is sandy) and (loam quality in soil) and (mountain Slope is high Altitude) then crop is **tea**.

Now considering the example of coconut which is not included in our database. It requires 1000 mm rainfall, coastal area with well-drained red, sandy loam soil with very high demand of NPK fertilizers. No such rule is present in our database. Also, the system has considered NPK up to high quantity only and not very high quantity as required by coconut. Hence, the system can be modified to include rules for other crops too. Coconut requires high water table, deep fertile soil and all these parameters are covered in the existing parameters.

Thus, we can see the proposed system works correctly for all the crops because the entire exhaustive list of parameters has been included.

6 Conclusion and Future Work

Although the proposed expert system at this stage is just a skeleton system sans any farmer friendly GUI, the input parameters chosen are exhaustive at least for Indian region, any number of new crop rules can be added to the system. In case the broad range of a linguistic variable causes conflict in two fuzzy rules, it can be narrowed down further. Hence, the proposed system is extremely customizable.

As the future work, the work can be extended to include more crops; but before that if a GUI-based market-ready front end for this system is designed, it will be a great step towards assisting the farmers. Furthermore, the expert system can be extended to include different clauses for crops in growing phase and in harvest period. These clauses can contain pest diagnosis [7] for each crop and harvesting tips too. The front end of the proposed system can also include details about seeds and varieties recommended by government for a crop thus increasing the system's credibility.

References

1. Ministry of Agriculture and Farmers Welfare, Official website of the Government of India
2. Ladha JK, Dawe D, Pathak H, Padre AT, Yadav RL, Singh Bijay, Singh Yadvinder et al (2003) How extensive are yield declines in long-term ricewheat experiments in Asia? Field Crops Res 81(2):159–180
3. Zadeh LA (1996) Fuzzy logic = computing with words. IEEE Trans Fuzzy Syst 4(2):103–111
4. Negnevitsky M (2005) Artificial intelligence: a guide to intelligent systems. Pearson Education, Harlow
5. Montalvo M, Guerrero JM, Romeo J, Emmi L, Guijarro M, Pajares G (2013) Automatic expert system for weeds/crops identification in images from maize fields. Expert Syst Appl 40 (1):75–82
6. Ponnusamy K, Sriram N, Prabhukumar S, Vadivel E, Venkatachalam R, Mohan B (2016) Effectiveness of cattle and buffalo expert system in knowledge management among the farmers. The Indian J Anim Sci 86(5)
7. Hasan SS, Solomon S, Baitha A, Singh MR, Sah AK, Kumar R, Shukla SK (2015) CaneDES: a web-based expert system for disorder diagnosis in sugarcane. Sugar Tech 17(4):418–427
8. Adekanmbi O, Green P (2014) A meta-heuristics based decision support system for optimal crop planning
9. Kawtrakul A, Amorntarant R, Chanlekha H (2015) Development of an expert system for personalized crop planning. In: Proceedings of the 7th international conference on management of computational and collective intelligence in Digital EcoSystems, ACM, pp 250–257
10. MathWorks, Inc., Wei-cheng W (1998) Fuzzy logic toolbox: for use with MATLAB: user's guide. Mathworks, Incorporated, Natick, MA
11. Dhaliwal HS, Kular JS (2014) Package of practices for the crops of Punjab. Punjab Agricultural University, Ludhiana
12. Joshi PK (2005) Maize in India: production systems, constraints, and research priorities. CIMMYT
13. Datta De (1981) Principles and practices of rice production. Int Rice Res Inst

14. Venugopalan MV, Sankaranarayanan K, Blaise D, Nalayini P, Prahraj CS, Gangaiah B (2009) Bt cotton (*Gossypium* sp.) in India and its agronomic requirements a review. Indian J Agron 54(4):343
15. Balezentiene Ligita, Streimikiene Dalia, Balezentis Tomas (2013) Fuzzy decision support methodology for sustainable energy crop selection. Renew Sustain Energy Rev 17:83–93
16. Jawad F, Choudhury TUR, Asif Sazed SM, Yasmin S, Rishva KI, Tamanna F, Rahman RM (2016) Analysis of optimum crop cultivation using fuzzy system. In: 2016 IEEE/ACIS 15th international conference on computer and information science (ICIS), IEEE, pp 1–6

Part V
Soft Computing Based Online Health Care Systems

Fuzzy Logic Based Web Application for Gynaecology Disease Diagnosis

A. S. Sardesai, P. W. Sambarey, V. V. Tekale Kulkarni,
A. W. Deshpande and V. S. Kharat

1 Introduction

In the process of medical diagnosis, generally, a clinician assigns a label to a patient based on the symptoms the clinician gathers from the patient and his/her medical knowledge. This assignment of label, the diagnosis involves several levels of uncertainty and imprecision which is too natural in medical field. The symptoms also overlap among different diseases. Also, several diseases present in a patient may interface with the usual description of any of the disease. A disease is noticeable in different patients quite differently depending upon the patient's physical state, which also shows different intensities per patient. The diseases can be precisely described using linguistic terms which are also imprecise and vague [15]. The diseases such as gynaecological ones involve various associated factors

A. S. Sardesai (✉)
Modern College of Arts, Science and Commerce, Shivajinagar,
Pune 411 005, India
e-mail: sardesaicompsci@moderncollegepune.edu.in

P. W. Sambarey
Department of Gynaecology and Obstetrics, B. J. Medical College, Pune, India
e-mail: drsambarey@yahoo.co.in

V. V. T. Kulkarni
Onshape India Pvt. Ltd., Pune, India
e-mail: vishnutekale13@gmail.com

A. W. Deshpande
Berkeley Initiative of Soft Computing (BISC)-SIG-EMS,
University of California, Oakland, USA
e-mail: ashok_deshpande@hotmail.com

V. S. Kharat
Department of Mathematics, Savitribai Phule Pune University, Pune, India
e-mail: laddoo1@yahoo.com

© Springer Nature Singapore Pte Ltd. 2017
R. Ali and M. M. S. Beg (eds.), *Applications of Soft Computing for the Web*,
https://doi.org/10.1007/978-981-10-7098-3_9

which are imprecise, vague and are difficult to represent using conventional statistical methods.

The words 'symptoms' and 'diagnosis' are used here very loosely to refer to the labels and the information, respectively. By 'symptoms', we mean any information obtained from the patient—in this particular case, only interviews. By 'diagnosis', we mean any label, any word used by the clinician to describe and synthesize the medical status of a patient. The tactic knowledge of the experts (perceptions) is needed for this synthesis. To summarize, the study is based on the patients and the perceptions of experts are invariably expressed in linguistic statements.

The medical knowledge is usually represented by symptom–disease relationship. This knowledge contains imprecision, uncertainty in two forms, viz. in the diagnosis process and the information narrated by the patient. The information gathered by the clinician from the patient is in the form of past history, physical examination, histopathology tests and other investigative tests. The information provided by the patient may be subjective, exaggerated underestimated or incomplete. The laboratory test results often are of limited precision. Moreover, all the investigative tests require correct interpretation of results which again show the vagueness [1, 33]. Considering such a type of uncertainty, vagueness in the parameters for process of diagnosis, it is difficult for the physician to diagnose the disease with correct precision. The use of fuzzy logic may overcome the difficulties in this diagnosis process. Additionally, consider a situation where a patient goes to a physician. The physician asks about complaints, patient tells symptoms like —'heavy pain in abdomen and a bit pain in lower abdomen'. What do we understand by the words 'a bit' or 'little'? Do they tell us exactly, how much the pain is in some numeric unit value? They do not. But the physician understands it by analysing the patient. The word which we use in day-to-day life without specifying numeric quantities shows the fuzziness. In sum, medicine is a fuzzy science!

The study reported in this sequel is centred only on the tactic knowledge of the identified eight gynaecologists and the linguistic descriptions of the symptoms by the patients about the systems.

It can be stated that there is inherent uncertainty seen in gynaecology in the form of vagueness/fuzziness. It calls for exploring the use of fuzzy set theoretic operations in medical decision-making. The desire to better understand and teach this difficult and important technique of medical diagnosis has prompted attempts to model the process with the use of fuzzy set theory [15]. The fuzzy set theory is characterized by its capability of expressing knowledge in a linguistic way, which allows to describe simple, human-friendly rules, known as interpretability. This gives the attractive reason and application of fuzzy set theory in medical diagnosis [28].

1.1 A Word on Gynaecology

Gynaecology or gynaecology (science of women) refers to the health of the female reproductive system such as uterus, vagina and ovaries, while obstetrics deals with care of women's reproductive tracts and their children during pregnancy (prenatal period), childbirth and the postnatal period. The research study presented in this sequel refers only to medical diagnosis in gynaecology.

The most common gynaecological symptom narrated by patients is 'Pain in abdomen', where the patient is usually unable to state the exact location of the abdomen as lower abdomen, pelvic region or abdomen. Another symptom is the case with the menstrual flow, which contains patient specific interpretation, vague nature of narration. For example, irregular menses, excessive menstrual flow and painful menses do not give any exact measure of any of these symptoms. They are interpreted by patients based on their individual level of thinking.

It is somewhat tricky for a gynaecologist to reach to a proper diagnosis because of the uncertainty/fuzziness/ambiguity present in the information received from patients and/or tests results. Another problem in the gynaecology disease diagnosis is increasing volume of patients. In such a scenario, the process of medical diagnosis is highly based on the medical knowledge of the clinician and his/her interpretation about the gathered symptoms and test results.

1.2 Differential Diagnosis

In the field of medicine, the differential diagnosis process is defined as the process by which a particular disease or condition is distinguished from others that present similar symptoms. This process is used by the clinicians to rule out the diseases which possess similar symptoms. In general, a differential diagnostic process is an efficient diagnostic method which is used to spot the presence of a disease entity where multiple alternatives are possible. This process uses evidences such as symptoms, patient history and medical knowledge to confirm the diagnostics [28].

In our study, we have tried to simulate this process of differential diagnosis. Figure 1 shows the details of the process.

1.3 Three-Stage Approach

In authors view, the gynaecological disease diagnosis can be an approach of three different stages as shown in Fig. 2. Stage 1, also referred to as initial screening, is based on the subjective information provided by patients to the clinician and it gives the output as a single or multiple disease diagnosis. Stage 2, in which *history* parameters past history, premenstrual changes, last menstrual period, marital status, parity, etc., are applied only if single disease diagnosis is not labeled for a patient in Stage 1. The patients who do not receive single diagnostic label at Stage 2 undergo investigative test like physical examination and various tests like ultra-sonography, X-ray, blood tests, etc., and are passed to Stage 3 [28] (Fig. 2).

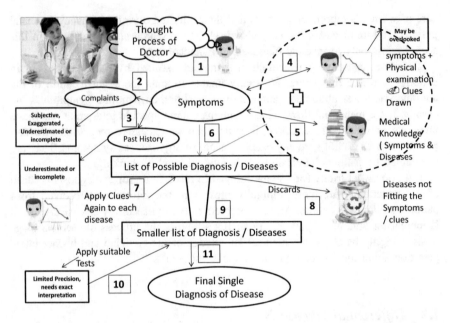

Fig. 1 Process of differential diagnosis

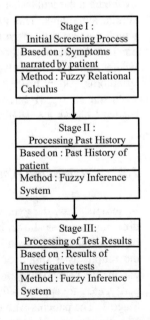

Fig. 2 Overall approach of
gynaecological disease
diagnosis

The objective of the study is to simulate the process of differential diagnosis using fuzzy logic based formalisms.

1.4 Web Application

The use of the software is well brought into play by the putting it on globe. The best way to do it is a development of web application. The application can be easily available over the net which will get accessed by people even at remote areas and can be used as a diagnosing tool for them. In our context, soft computing in web, it should be a web application which has the ability to solve lot of complex problems in different areas like fuzzy logic, neural network, etc. Web application is a client–server-based application in which the client (or user interface) runs in a web browser.

2 Literature Survey

Prof. Lotfi A. Zadeh published 'Fuzzy Sets' (Zadeh 1965), a landmark paper that gave the field its name. Prof. Zadeh applied Lukasiewicz logic to every object in a set. He proposed a complete algebra for fuzzy sets.

Uncertainty is a challenging part in human's everyday life. The main cause of uncertainty is the information deficiency. Information may be incomplete, fragmentary, not fully reliable, vague, contradictory or deficient in some other way. These various information deficiencies may result in different types of uncertainty. Seising [30] stated that the history of philosophy and medical diagnosis is concerned with the phenomenon of vagueness in the physician's 'style of thinking'. When physicians make a diagnosis, the use of fuzzy sets and fuzzy systems to create a model of such reasoning can be well suited.

2.1 Health Care and Medicine Using Fuzzy Logic

The development of expert systems in medical diagnosis has been started since 1970. The pioneer of medical expert systems, Sanchez [21–23] developed diagnosis systems using fuzzy relations representing relationships between symptoms and diseases. His work was extended by Adlassnig [4, 5, 8] using the relation between disease and symptoms in the form of occurrence and confirmability relations [17].

A Medical School in University of Vienna contributed pioneered research developments in the medical diagnosis [2, 3, 6–9, 20].

Steimann [16, 31, 32], Innocent [13, 14], Torres and Nieto [33] and Seising [30] analysed the evolution of fuzzy logic for various medical applications and diagnosis of variety of diseases. In a nutshell, fuzzy logic is being used since last four decades in medical diagnostics in almost all the branches [29].

2.2 Medical Diagnosis Using Fuzzy Expert Systems

There are various techniques in which the research in the field of medical expert systems has been carried out such as case-based reasoning, rule-based reasoning, artificial intelligence, inference relations, uncertainty modeling, web-based medical and decision support system [12].

Using these techniques, many different medical diagnostic systems were developed, some of which are still in use. The details of these systems are as follows [28, 29].

1. INTERNEST-I: Rule-based expert system for general internal medicine. A tree structure database which related diseases with symptoms.
2. MYCIN: The first well-known rule-based medical expert system for meningitis and bacteremia infections. Uses backward chaining reasoning strategy.
3. CADUCEUS: A medical expert system to improve MYCIN. CADUCEUS eventually could diagnose 1000 diseases.
4. QMR—Quick Medical Reference: Developed from INTERNIST-I, provides electronic access to more than 750 diseases. Uses more than 5000 clinical findings to describe the features of diseases in the QMR knowledge base.
5. PUFF—Pulmonary Function System: Expert system to diagnose the results of pulmonary function tests. Initially developed on the SUMEX computer using E-MYSCIN.
6. ATHENA: Support blood pressure control and suggests guideline-concordant choice of drug therapy. Clinical experts can customize the knowledge base so as to include new evidence.
7. CEMS: The mental health decision support system. The input to CEMS is clinical functions like evaluation, diagnosis, treatment and outcome assessment by allowing queries.
8. ERA—Early Referrals Application: Web-based decision support for cancer referrals. ERA is an interactive decision support tool to support general practitioners.
9. Iliad: Expert system for internal medical diagnosis. Uses Bayesian reasoning to calculate the posterior probabilities of various diagnoses under consideration.
10. GIDEON: Diagnosis of infectious and related diseases using Bayesian ranked differential diagnosis.
11. PERFEX: Objective: to develop a clinically useful, computer-based methodology to use in the diagnosis of heart disease. Contains well established mathematical methods, visualization techniques and artificial intelligence approaches.
12. ISABEL: Web-based diagnosis decision support system interfaced with Electronic Medical Record System. Covers all ages (neonates to geriatrics) and all major specialties and sub-specialties and is fast and easy to use.

13. Doctor Moon: Fuzzy logic rule-based system for diagnosis of lung diseases. Easy to install and has a friendly interface. The knowledge base of Doctor Moon is managed by a Borland Paradox Database consisting of 700 records, each represents a rule.
14. LISA: The Clinical information system for childhood acute lymphoblastic leukemia using a centralized Oracle database.
15. MEDUSA: Fuzzy expert system for medical diagnosis of acute abdominal pain. It uses rule-based, heuristic and case-based reasoning on the basis of imprecise information.
16. CADIAG-2: Computer Assisted DIAGnosis is a well-known rule-based expert system for internal medicine containing the inference engine and the knowledge base based on probabilistic logic theory.
17. MedFrame/CADIAG IV: Rule-based and frame-based diagnostic and remedial consultation system. It is multiuser Client–Server GUI system.
18. Medical Diagnostic System: An expert system used to support the diagnosis and treatment of diabetes. Logic-based system containing forward and backward chaining inference mechanism which is based on evidence theory.
19. Dr. Watson: Experts system developed by IBM with natural language processing, hypothesis generation and evidence-based learning. Parses the input to identify the key pieces of information and then mines the patient data to find relevant facts about family history, current medications and other existing conditions.
20. Fuzzy Logic Based Gynaecology Disease Diagnosis System: The authors have devised a Fuzzy Logic Based Gynaecology Disease Diagnosis System which simulates the process of differential Diagnosis [26, 27].

3 Techniques Used

3.1 Fuzzy Set Theory and Fuzzy Logic

Computers can understand only true and false values whereas human being can interpret the degree of truth or degree of falseness. These human actions can be interpreted by fuzzy logic and is called as machine learning or artificial intelligence.

3.1.1 Fuzzy Sets

The father of fuzzy set theory is an American Professor *A. Zadeh.* In 1965 when he presented the seminal paper on fuzzy sets, he showed that fuzzy logic unlike classical logic can realize values between 0 (false) and 1 (true). He transformed the crisp (classical) set into the continuous set. The fuzzy sets have movable boundaries, i.e. elements of fuzzy set represent true false value and also represent the degree of truth or degree of falseness for each input [19, 24].

Definition:

a. A fuzzy set A is defined by an ordered pair, a binary relation

$$A = \{(x, \mu_A(x)) | x \in X, \quad \mu_A(x) \in [0, 1]\},$$

where $\mu_A(x)$ is a membership function which specifies grade or degree to which any element x in A belongs to the fuzzy set A.

A is a fuzzy set, X is universe or universe of discourse.

Elements with $0°$ of membership in a fuzzy set are usually not listed in the fuzzy set.

b. If the universe of discourse X is **discrete and finite**, Fuzzy set A is represented as follows.

$$A = \left\{ \frac{\mu_A(x_1)}{x_1} + \frac{\mu_A(x_2)}{x_2} + \cdots \right\}$$

c. If the universe of discourse X is **continuous and infinite**, Fuzzy set A is represented as follows:

$$A = \left\{ \int \frac{\mu_A(x)}{x} \right\}$$

In both the above notations, the horizontal bar is not quotient but it is a delimiter where denominator → element of universe.

Numerator in each term → the membership value in set A associated with the element of the universe indicated in the denominator.

The summation symbol ($\sum$) is not algebraic summation, it denotes collection or aggregation of each element. The '+' sign is not algebraic 'addition' but it denotes collection or aggregation of each element. The integral sign ($\int$) is not algebraic integral but it is a continuous function-theoretic aggregation operator for continuous variables.

3.1.2 Fuzzy Relational Calculus (Compositional Rule of Inference)

The definitions of fuzzy sets and fuzzy operations are well known and therefore are not discussed in this section. This section very briefly describes fuzzy relational calculus used in the study [11, 15, 19, 27].

A fuzzy relation R maps the elements from universe X to universe Y and S is the relation that maps the elements from universe Y to Z. The membership functions $\mu_R(x, y)$ and $\mu_S(y, z)$ give the strength of the mapping. Fuzzy max–min composition is defined in terms of the function-theoretic notation in the following manner:

$$\mu_T(x, z) = \vee_{y \in Y}(\mu_R(x, y) \wedge (\mu_S(y, z))) \tag{1}$$

Definition 1 Let $R \subset X \times Y$ and $S \subset Y \times Z$ be fuzzy relations, the max–min composition
$R \circ S$ is defined by

$$R \circ S = \{((x, z), \max_y\{\min_{x,z}\{f_R(x, y), f_S(y, z)\}\}) | x \in X, y \in Y, z \in Z\} \tag{2}$$

3.1.3 Type 1 Fuzzy Inference System (T1FIS)

Fuzzy inference is the process of establishing the mapping from a given input to an output using Fuzzy Logic. The process of fuzzy inference involves fuzzification, defuzzification, implication and aggregation.

Fuzzification:

Fuzzification is a process of changing real scalar value into a fuzzy value. This is achieved with the different types of fuzzifiers or different types of membership functions, viz. Singleton fuzzifier, triangular membership function, trapezoidal membership function, Gaussian membership function, bell-shaped membership function, Z-shaped membership function, sigmoidal membership function, S-shaped membership function, π_1 membership function and π_2 membership function. In the research, trapezoidal membership function is used.

3.1.4 Trapezoidal Membership Function

By a trapezoidal membership function, we mean the regular uncertain set is fully determined by a quadruplet (a, b, c, d) of crisp numbers with $a < b < c < d$ whose membership function is shown in Eq. (3).

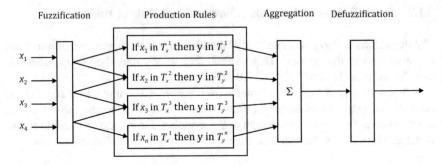

Fig. 3 Schematic diagram of FIS

$$\mu_A = x \begin{cases} 0 & x \leq \alpha \\ (x - \alpha)/(\beta - \alpha), & \alpha \leq x \leq \beta \\ 1 & \beta \leq x \leq \gamma \\ (\delta - x)/\delta - \gamma), & \gamma \leq x \leq \delta \\ 0 & x \geq \delta \end{cases} \quad (3)$$

The linguistic variable x_i in U is characterized by

$$T(x) = \{Tx^1, Tx^2, Tx^3, \ldots, Tx^k\} \quad (4)$$

and

$$\mu(x) = \{\mu x^1, \mu x^2, \mu x^3, \ldots, \mu x^k\} \quad (5)$$

where $T(x)$ is a term set of x. The linguistic variable y in V is characterized by

$$T(y) = \{Ty^1, Ty^2, Ty^3, \ldots, Ty^k\}, \quad (6)$$

where $T(y)$ is a term set of y, i.e. T *is* the set of names of linguistic values of y, with each Ty^i being a fuzzy membership function μx^i defined on V [19, 24, 26] (Fig. 3).

Mamdani-type fuzzy inference system is used in this research which uses centre of gravity as defuzzification method. Software is developed to simulate the Mamdani-type fuzzy inference system.

3.1.5 Defuzzification

Defuzzification is a method used to convert the fuzzy quantity to crisp quantity, which is exactly opposite to fuzzification which converts crisp value to fuzzy value. Some of the defuzzification methods are Max membership principle, centroid method, weighted average method, mean max membership, center of sums, center of largest area and first (or last of maxima). Center of gravity defuzzification method is used in the study.

3.2 The Web Technology

Nowadays, many big web applications are getting implemented using N-Tier architecture and REST services. REST services are common interface for the application business logic. This gives the benefits of writing business logic once and any consumer who can communicate with these services can simply request and get the desired output in response of that service. The interoperability between computer systems on the Internet is provided by RESTful (Representational state transfer) web services. Through REST-compliant web services, the textual representation web resources are allowed which use a predefined and uniform set of stateless operations.

This web application is a RESTfull [18] application which has a thin UI client serving multiple devices like desktop, tablet and smart phones. End user accesses this application over web with their device and they submit their problem details to the application in user-friendly way.

The diagnosis algorithm is executed on the remote server. This helps in heavy lifting of algorithms execution and will share the result as a response to the REST API to the client. The web application is hosted on the web server which uses web service.

Working

Client browser makes an AJAX request to web server with the posted data depending on the query to the application. Once the request is received at the web server, it communicates with application server. Application server is one which is responsible for the execution of core logic of the computing. Once the processing is completed, application server sends the message to web server. Web server informs client about the completion of the processing by either push notification, polling from client. The working is as shown in Fig. 4.

Following layers are used in REST:

1. *Client Layer:* There are many popular JavaScript frameworks in this segment like AngularJs, ReactJs, EmberJs, etc.

   Client layer will be combination of any of the JS framework, HTML5 and CSS3. This will give us freedom to target any devices which can run webapp.

2. *Server layer:* We can use any existing server side technology to write the REST services like Java, .NET, PHP, NodeJs, etc.
3. *Application Layer:* Same like server layer. This will mostly work as rule engine and contain algorithms for solving some problems.

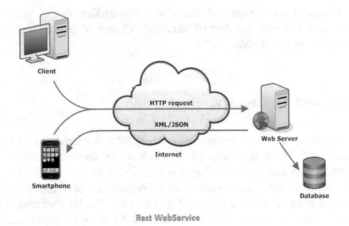

Fig. 4 REST WebService

4. *DB layer:* If application needs database, we can use currently available open source databases like mysql, mongoDB, etc.

3.2.1 REST API Design

The services based on REST technology have their own task of sending input or live data depending on provided input. This hides all the complexity of generating this data from consumer and provides the abstract implementation where consumer can call and take the response. REST service communicates with other service to get some other data. There are six guiding constraints restricting the way that the server responds to the client requests or its process. Based on these constraints the desirable, nonfunctional properties such as simplicity, performance, reliability, visibility, modifiability, scalability and portability are gained. The service violating any of the above-mentioned constraints is not RESTful.

REST API design is as below. Every entity exposes GET, POST, PUT, DELETE APIs for consumers [10] (Figs. 5 and 6).

Here, the client sends the symptoms backache and fever to the server in the specified format as request parameter and the server sends the response as the name of the disease flu. The use of REST in the application has lead the application to be categorized in SASS (Software as Service). Here all the core logic, business logic or in our case diagnostic logic is executed at server tier. The server communicates to client via JSON response.

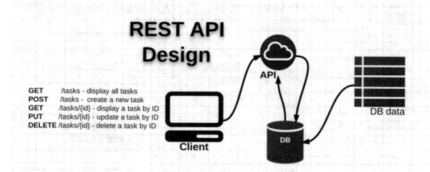

Fig. 5 REST API design

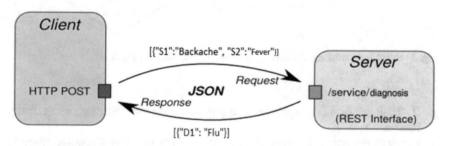

Fig. 6 JOSN/REST/HTTP

3.3 Three-Stage Approach

In our view, overall approach in gynaecological disease diagnosis could be as shown in Fig. 7 [28].

a. Get the perceptions from experts and store in two different flat files.
b. Collect patients data by interview method in the form of signs and symptoms. Signs indicate the history of the patient, i.e. Age, PMC, LMP, Parity, etc. Symptoms indicate the complaints narrated by the patient.
c. Store the patient data in a flat file structure. Use of flat file has given a fast access of data, reducing time complexity thereby.
d. Apply Max-min composition method to get the Indication matrices.
e. Diagnose a disease. Check if a single disease is diagnosed. If yes, it is the final diagnosis. Stage I (Initial Screening) and the diagnosis process end here.
f. If single disease is not diagnosed, we continue the diagnosis process. Check if the 'History' factor is applicable to the patient.

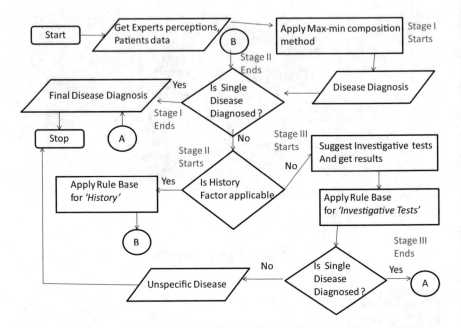

Fig. 7 Broad level working of the proposed model is as shown in Fig. 7

g. If 'History' factor is applicable, apply rule base for 'History' using fuzzy inference system.
h. Check if the output is single disease. If yes, it is the final diagnosis. Stage II (History Rule base) and the diagnosis process end here.
i. If the output is not a single disease at Stage II also, then further investigative tests are suggested and the test results are obtained.
j. Test rule base is applied on the test results to get the final diagnosis.
k. If the single disease is obtained after application of test rule base, it is the final diagnosis. Stage III (Test Rule base) and the diagnosis process end here.
l. If at Stage III also we do not get the desired output, the model gives output as 'Unspecific Disease'.

4 The Method

4.1 Perception-Based Modeling

Eight different gynaecologists' perceptions were collected from Pune, Ahmednagar, Ambejogai and Pimpri. These perceptions are used as master input data for the initial screening stage to get the diagnosis at Stage I. This is called experts' knowledgebase.

These are the perceptions of the experts developed over more than 30 years of their experience in the fields of gynaecology. We modeled these perceptions to diagnose a patient in the initial screening stage.

A. Stage I (Initial Screening)

The computational framework proposes two types of fuzzy relations which exist between symptoms (s) and diseases (d). These are fuzzy occurrence relation (R_o) and fuzzy confirmability relation (R_c). The first gives knowledge about the frequency of a symptom when the specific disease is present; it corresponds to the question

How often does symptom, 's' occur with disease 'd'?

The second fuzzy relation describes perceptive power of the symptom to confirm disease; it corresponds to the question

How strongly does symptom, 's' confirm disease 'd'?.

The distinction between occurrence and confirmability is useful because a symptom may often occur in given disease but the presence of this symptom may not confirm the disease whereas the presence of the same symptom in some other disease may confirm the disease. On the other hand, another symptom may occur rarely but the presence of this symptom confirms the presence of disease.

During Stage I, we collected eight experts' perceptions for symptom–disease relationship. Symptoms were recorded for 226 gynaecology patients using personal interview technique from three different hospitals in Pune, Pimpri and Ambejogai. We revisited fuzzy relational calculus and developed the software of disease diagnosis [25, 27]. The results obtained by the software are verified with the actual diagnosis given by the gynaecologists [26, 28].

B. Stage II (History Rule Base)

The diagnosis in gynaecology depends on the history of the patient in terms of patient's menses details, marital status, number of children, etc. Based on these factors, five fuzzy set are defined as:

   i. *'Age'*
  ii. *'LMP'*
 iii. *'PMC_Flow'*
 iv. *'Marital Status'*
  v. *'Parity'*

Based on these fuzzy sets, 72 fuzzy rules are formulated and a model is developed using type 1 fuzzy inference system [26].

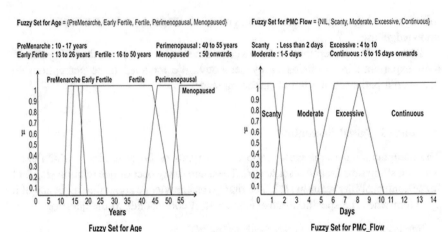

Fig. 8 Fuzzy sets for age and PMC_Flow

Figure 8 shows two fuzzy sets '*Age*' and '*PMC_Flow*'.

Figure 9 demonstrates the application of fuzzy rules using Mamdani-type fuzzy inference system, for patient (P_{79}) with history as—Age—34, Parity = 1, M_Status = Married 4 years back, LMP 60 days back. These three factors will fire following 12 rules from the set of 72 rules to yield the output as: '*Incomplete Abortion present with possibility 1*'.

C. **Stage III (Test Rule Base)**

If single disease is not diagnosed at stage II, the investigative tests are suggested some of which are as shown in following table.

Disease	Tests
Endometriosis	Transvaginal USG
	Pelvic laparoscopy
Vaginal yeast infection	KOH test
Ovarian cyst	USG pelvis
	CT scan
	Doppler flow studies
	MRI ovary
	Hormone levels—estradiol, testosterone, HCG
	Serum HCG
Urinary tract infection	Urinalysis/microscopic
	Urine culture
	CBC

(continued)

(continued)

Disease	Tests
Ovarian cancer	CA-125 blood test (1)
	Complete blood count & blood chemistry
	CT scan or MRI of pelvis or abdomen
	USG pelvis
	Pelvic laparoscopy (for small lumps)
	Exploratory laparotomy
	Biopsy
Pelvic inflammatory disease (PID)	CRP: c-reactive Protein
	Erythrocyte sedimentation rate (ESR)
	WBC count
	Culture of vagina or cervix
	Pelvic USG
	CT scan
	Laparoscopy
Dysfunctional uterine bleeding (DUB)	Blood clotting profile
	Hormone tests (FSH, LH)
	Progesterone
	Endometrial biopsy
	Transvaginal USG
Uterine fibroid	USG pelvis
	CBC
	MRI
Endometritis	USG transvaginal
	MRI

The results of these investigative test are input to stage III which is a rule base designed for 13 diseases and related imagining, pathology tests and physical examinations. Fuzzy sets are defined after consulting radiologists, pathologists and gynaecologists.

For example, fuzzy set for USG_Pelvis_Ovarian_Cyst (absent, small, medium, large) is as shown below:

We have defined 55 fuzzy sets for investigative tests and are confirmed by six experts. For these 55 fuzzy sets, we defined 715 fuzzy rules.

For example, to test if the '*Ovarian Cyst*' is present, one of the tests suggested is '*USG Pelvis*'. Figure 10 shows the fuzzy set defined for this test.

For patient P_{140}, the diagnosis by Stage I is {*Pelvic Inflammatory Disease (PID), Endometritis, Ovarian Cyst*}, Stage II cannot arrive at final diagnosis so investigative test is suggested '*USG Pelvis*' which resulted in cyst present of size 7.6 cm (Large cyst). So the fuzzy rule fired is as shown in Fig. 11.

The patients who could not get diagnosed by stage II are input to stage III. Out of 226 patients, 50 patients were diagnosed for a single disease and for 11 patients

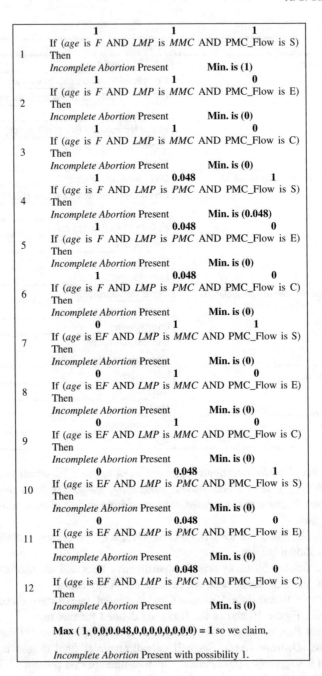

	1 **1** **1**
1	If (*age* is *F* AND *LMP* is *MMC* AND PMC_Flow is S) Then *Incomplete Abortion* Present **Min. is (1)**
2	**1** **1** **0** If (*age* is *F* AND *LMP* is *MMC* AND PMC_Flow is E) Then *Incomplete Abortion* Present **Min. is (0)**
3	**1** **1** **0** If (*age* is *F* AND *LMP* is *MMC* AND PMC_Flow is C) Then *Incomplete Abortion* Present **Min. is (0)**
4	**1** **0.048** **1** If (*age* is *F* AND *LMP* is *PMC* AND PMC_Flow is S) Then *Incomplete Abortion* Present **Min. is (0.048)**
5	**1** **0.048** **0** If (*age* is *F* AND *LMP* is *PMC* AND PMC_Flow is E) Then *Incomplete Abortion* Present **Min. is (0)**
6	**1** **0.048** **0** If (*age* is *F* AND *LMP* is *PMC* AND PMC_Flow is C) Then *Incomplete Abortion* Present **Min. is (0)**
7	**0** **1** **1** If (*age* is E*F* AND *LMP* is *MMC* AND PMC_Flow is S) Then *Incomplete Abortion* Present **Min. is (0)**
8	**0** **1** **0** If (*age* is E*F* AND *LMP* is *MMC* AND PMC_Flow is E) Then *Incomplete Abortion* Present **Min. is (0)**
9	**0** **1** **0** If (*age* is E*F* AND *LMP* is *MMC* AND PMC_Flow is C) Then *Incomplete Abortion* Present **Min. is (0)**
10	**0** **0.048** **1** If (*age* is E*F* AND *LMP* is *PMC* AND PMC_Flow is S) Then *Incomplete Abortion* Present **Min. is (0)**
11	**0** **0.048** **0** If (*age* is E*F* AND *LMP* is *PMC* AND PMC_Flow is E) Then *Incomplete Abortion* Present **Min. is (0)**
12	**0** **0.048** **0** If (*age* is E*F* AND *LMP* is *PMC* AND PMC_Flow is C) Then *Incomplete Abortion* Present **Min. is (0)**

Max (1, 0,0,0.048,0,0,0,0,0,0,0,0) = 1 so we claim,

Incomplete Abortion Present with possibility 1.

Fig. 9 Fuzzy rules for patient (P_{79})

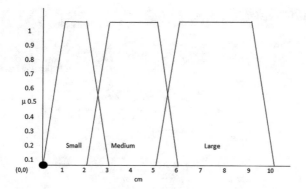

Fig. 10 Fuzzy set for USG_ PELVIS_OVARIAN_CYST

(0)   **(0)**
if(*USGPV_Ocyst* is SMALL OR *USGPV_Ocyst* is MEDIUM
(1)
OR *USGPV_Ocyst* is LARGE)
(1)
then *Ovarian Cyst* is Present

Fig. 11 Fuzzy rule fired for patient P_{140}

the diagnosis by model was '*Disease Unspecific*'. Stage II diagnosed more 29 patients; remaining 138 patients are input to stage III.

The GUI of the website is as shown in Fig. 12.

5 Results and Discussion

Stage I:

Table 1 shows that max–min composition does not result into only single disease diagnosis but identifies more than one disease or multiple diagnoses in a patient which is termed as initial screening by the physicians. The output of the model is same as the diagnosis by physician at this stage. For patient P_9, the model has diagnosed a single disease: Pelvic Inflammatory Disease (PID). For all other seven patients, the gynaecologist suggests further investigative tests to arrive at the single disease diagnosis. In patient P_8, the diagnosis given by model is '*Disease Unspecific*'. There are various reasons to receive output as '*Disease Unspecific*'. Some of these reasons could be as follows:

a. The symptoms narrated by the patient may not match with the symptoms of any of the currently selected 31 diseases.

Fig. 12 Screenshot of the web application

Table 1 Diagnosis of patients in case study example (case study example)

Patient	Possible diseases occurred in the patient
P_1	Uterine Fibroid (D_3), Adenomyosis (D_4), Endometriosis (D_7), Pelvic Inflammatory Disease (D_8)
P_2	Vaginal Yeast Infection (D_1), Ovarian Cyst (D_2), Cervicitis (D_6), Pelvic Inflammatory Disease (D_8), Leucorrhoea (D_9)
P_3	Ovarian Cyst (D_2), Pelvic Inflammatory Disease (D_8)
P_4	Vaginal Yeast Infection (D_1), Cervicitis (D_6), Leucorrhoea (D_9)
P_5	Uterine Fibroid (D_3), Adenomyosis (D_4), Endometriosis (D_7), Pelvic Inflammatory Disease (D_8)
P_6	Vaginal Yeast Infection (D_1), Cervicitis (D_6), Endometriosis (D_7), Leucorrhoea (D_9)
P_7	Uterine Fibroid (D_3), Adenomyosis (D_4), Dysfunctional Uterine Bleeding (D_5)
P_8	Disease Unspecific
P_9	Pelvic Inflammatory Disease (D_8)

b. The symptoms with specific severity narrated by the patients may not result in a possibility value which is decided as a α-cut value in the current model.

c. It is mostly observed in many patients that the patient with psychic history cannot be diagnosed correctly.

An overview of the diagnosis analysis for one expert using initial screening stage is presented below.

Table 2 Patients diagnosed using Stage II

Patient id	Diagnosis
P_{164}	Secondary amenorrhea
P_{22}, P_{103}, P_{215}	UPT to be done
P_{100}	No diagnosis
P_{90}, P_{226}	Incomplete abortion
P_{20}, P_{93}, P_{154}, P_{192}, P_{199}, P_{200}, P_{225}	Secondary amenorrhea (UPT to be done)
P_{52}, P_{69}, P_{218}	Cervical cancer
P_{129}	Cervical cancer/DUB
P_{79}	Incomplete abortion/DUB
P_{187}	Pubertal menorrhagia
P_{153}, P_{191}, P_{223}	Endometriosis/fibroid
P_{54}, P_{65}, P_{138}	DUB
P_{132}, P_{137}, P_{202}	Perimenopausal
P_{14}	Uterine cancer

Table 3 Diagnosis of Stage III over the input from Stage I and Stage II diagnosis (Sample cases)

Patient Id	Diagnosis by model (Stage I and II)	Diagnosis by model (Stage III)
P_{46}	Uterine prolapse, Urinary tract infection	Uterine prolapse
P_{49}	Uterine prolapse, Vaginal Yeast Infection	Uterine prolapse
P_{79}	Dysfunctional Uterine Bleeding (DUB), Abortion	Abortion
P_{89}	Disease Unspecific	Polycystic Ovarian Syndrome
P_{123}	Endometriosis, Uterine Fibroid, Adenomyosis, Pubertal menorrhagia	Pubertal Menorrhagia
P_{129}	Dysfunctional Uterine Bleeding (DUB), Cervical Cancer	Dysfunctional Uterine Bleeding
P_{153}	Endometriosis, Uterine Fibroid	Endometriosis
P_{191}	Endometriosis, Uterine Fibroid	Endometriosis
P_{194}	Secondary amenorrhea,	Polycystic Ovarian Syndrome
P_{223}	Endometriosis, Uterine Fibroid	Endometriosis
P_{224}	Endometriosis, Dysfunctional Uterine Bleeding(DUB), Uterine Fibroid, Adenomyosis, Pubertal Menorrhagia	Pubertal Menorrhagia

- Total number of patients diagnosed by model = 226.
- Number of patients with only one disease as diagnosed by model and confirmed by the expert = 50.
- Number of patients with multiple diseases as diagnosed by model, all correct and confirmed by the expert = 50.

Table 4 Output of differential diagnosis process

	Stage I	Stage II	Stage III
Number of patients input to stage	226	30	152
Number of patients correctly diagnosed as single disease	50	24	141
Number of patients incorrectly diagnosed	29	1	11
Accuracy percentage (%)	22.12	80.00	92.81
Overall accuracy of the model (%)	95.13		

- Number of patients with multiple diseases as diagnosed by model, partial correct and confirmed by the expert = 107.
- Diagnosis percentage with 100% accuracy by the model (Only 1 disease is diagnosed by model) and confirmed by the expert = 22.12%.
- Diagnosis percentage of initial screening model diagnosis with 1 + multiple diseases confirmed by the expert = 87.17%.
- Failure percentage of initial screening model = 12.83%.

Stage II:

Stage I gives correct diagnosis of 50 patients. Out of remaining 176 patients, Stage II identified 30 patients for the '*history*' criteria. It was found that 29 out of 30 patients are correctly diagnosed [26] (Table 2).

Stage III: *Tests* Inference System with Fuzzy Rule Base

Out of 176 patients, 29 patients are correctly diagnosed in Stage II. Out of these 29 patients, 5 patients (P_{79}, P_{153}, P_{191}, P_{129}, and P_{224}) are required to be passed to Stage III as they have received multiple disease diagnosis. In short, 152 patients need assistance of Stage III. Depending upon the symptoms they possess and their diagnosis done by initial screening stage, the investigative tests are suggested to these 152 patients. The respective test results are given to the Stage III.

Table 3 shows some of the cases along with the details of their diagnosis by Stage I and II and the final diagnosis by Stage III.

The summary of the overall diagnosis process for three stages is as given in Table 4.

6 Conclusion

The development of medical decision support system (MDSS) addresses perception-based diagnosis of patients for gynaecological diseases. This method might be suitable for other MDSS as well, with suitable modifications.

Many a times, a general comment made by some of the critics is 'Domain experts do not agree with one another'. The FST-based approach, coupled with a defined statistical method, should be used in final expert selection before embarking on application of fuzzy sets and fuzzy logic system.

The commonly seen benefits of this web application over the existing desktop application can be listed as below.

1. Web applications avoid the burden in deploying in each client machine.
2. Can access from anywhere.
3. Platform independent.
4. Updates are easier. No installation or upgrade patch required.
5. Do not have to enforce version check in client machine.
6. Makes bug fixes easier.
7. No administrator rights checking.
8. Support and maintenance are easier.
9. Adaptability in mobile applications.

In the context of our application, our end users who mostly will be ladies can easily access the website instead of installing a desktop application.

The use of web application can be thought of a useful tool to reach people at remote areas especially the areas which lack medical facilities.

References

1. Abbod MF, von Keyserlingk FG, Linkens DA, Mahfouf M (2001) Survey of utilisation of fuzzy technology in medicine and healthcare. Fuzzy Sets Syst 120:331–349
2. Adlassnig KP, Kolarz G, Scheithauer W, Effenberger H, Grabner H (1985) CADIAG: approaches to computer-assisted medical diagnosis. Comput Biol Med 15:315–335
3. Adlassnig KP, Kolarz G, Scheithauer W, Grabner H (1986) A approach to a hospital-based application of a medical expert system. Med Inform 11:205–223
4. Adlassnig KP (1980) A fuzzy logical model of computer assisted medical diagnosis. Methods Inf Med 19:141–148
5. Adlassnig KP (1982) A survey on medical diagnosis and fuzzy subsets. In: Gupta MM, Sanchez E (eds) Approximate reasoning in decision analysis, North-Holland Publishing Company, Amsterdam, pp 203–217
6. Adlassnig KP (1988) Uniform representation of vagueness and impression in patient's medical findings using fuzzy sets. In: Trappl R (ed) Cybernatics and systems. Kluwer Academic Publishers, Dordrecht, pp 685–692
7. Adlassnig KP (2001) Fuzzy set theory and fuzzy logic in medicine. In: Proceedings of 10th international conference system-modeling-control, Poland
8. Adlassnig KP (1986) Fuzzy set theory in medical diagnosis. IEEE Trans Syst Man Cybern 16 (2):260–265
9. Brein L, Adlassnig KP, Kolousek G (1998) Rule base and inference process of the medical expert system CADIAG-IV. In: Trappl R (ed) Cybernetics and systems'98, Vienna, Schottengasse, 3, A-1010, Austrian Society of Cybernetic Studies, pp 155–159
10. csc530 (2017) http://rfka.tripod.com/csc530.htm
11. Fuller R, Zimmermann HJ (1993) On Zadeh's compositional rule of inference. Lowen R, Roubens M (eds) Fuzzy logic: state of the art, theory and decision library, series D, Kluwer Academic Publisher, Dordrecht, pp 193–200. ISBN 0-7923-2324-6
12. Gulavani SS, Kulkarni RV (2009) A review of knowledge based systems in medical diagnosis. Int J Inf Technol Knowl Manag 2(2):269–275
13. Innocent PR, John RI (2004) Computer aided fuzzy medical diagnosis. Inf Sci 162:81–104

14. Innocent PR, John RI, Garibaldi J (2004) Fuzzy methods and medical diagnosis. Inf Sci 162:81–104
15. Klir GJ, Yuan CB (1995) Fuzzy sets and fuzzy logic, theory and applications. Prentice Hall P.T.R., Upper Saddle River, New Jersey
16. Kuncheva LI, Steimann F (1999) Fuzzy diagnosis. Artif Intell Med 16(2):121–128
17. Mahdi AA, Razali AM, Salih AA (2011) The diagnosis of chicken pox and measles using fuzzy relations. J Basic Appl Sci Res 1(7):679–686
18. Representational_state_transfer (2017) Received from https://en.wikipedia.org/wiki/Representational_state_transfer
19. Ross TJ (1995) Fuzzy logic with engineering applications. Willy, The Atrium Southern Gate, Chichester, West Sussex, United Kingdom
20. Sageder B, Boegl K, Adlassnig KP, Kolousek G, Trummer B (1997) The knowledge model of MedFrame/CADIAG-IV. Medical informatics Europe'97, pp 629–633
21. Sanchez E (1976) Resolution of composite fuzzy relation equations. Inf Control 30:38–48
22. Sanchez E (1977) Solutions in composite fuzzy relation equation. Application to medical diagnosis in Brouwerian logic. In: Gupta MM, Saridis GN, Gaines BR (eds) Fuzzy automata and decision process, Elsevier, North-Holland
23. Sanchez E (1979) Inverse of fuzzy relations, application to possibility distributions and medical diagnosis. Fuzzy Sets Syst 2(1):75–86
24. Sardesai A (2012) A text book of soft computing. Vision Publications, Pune. ISBN 978-93-5016-147-0
25. Sardesai A, Khrat V, Deshpande A, Sambarey P (2012) Initial screening of gynecological diseases in a patient, expert's knowledgebase and fuzzy set theory: a case study in India. In: Proceedings of 2nd world conference on soft computing, Baku, Azerbaijan, pp 258–262
26. Sardesai A, Khrat V, Deshpande A, Sambarey P (2014) Fuzzy logic application in gynaecology: a case study. In Proceedings of 3rd international conference on informatics, electronics and vision 2014, Dhaka, Bangladesh, IEEE explore digital library, 2014, pp 1–5. ISBN 978-1-4799-5179-6
27. Sardesai A, Khrat V, Deshpande A, Sambarey P (2016) Fuzzy logic based formalisms for gynaecology disease diagnosis. J Intell Syst, 283–295. https://doi.org/10.1515/jisys-2015-0106
28. Sardesai A, Khrat V, Deshpande A, Sambarey P (2017a) Fuzzy logic via computing with words in gynaecology disease. Dissertation, Savitribai Phule Pune Univeristy
29. Sardesai A, Khrat V, Deshpande A, Sambarey P (2017b) Artificial intelligence in medicine: a review. Int J Innov Comput Sci Eng, pp 19–27. http://ijicse.in/wp-content/uploads/2017/02/SOUVENIR-WWW.IJICSE.IN_.pdf
30. Seising R (2006) From vagueness in medical thought to the foundations of fuzzy reasoning in medical diagnosis. Artif Intell Med 38:237–256
31. Steimann F (1997) Fuzzy set theory in medicine. Artif Intell Med 11:1–7
32. Steimann F (2001) On the use and usefulness of fuzzy sets in medical AI. Artif Intell Med 21:131–137
33. Torres A, Nieto JJ (2006) Fuzzy logic in medicine and bioinformatics. J Biomed Biotechnol 2006. Article ID 91908, 1–7. https://doi.org/10.1155/JBB/2006/91908

Part VI
Soft Computing Based Online Documents Clustering

An Improved Clustering Method for Text Documents Using Neutrosophic Logic

**Nadeem Akhtar, Mohammad Naved Qureshi
and Mohd Vasim Ahamad**

1 Introduction

As a technique of Information Retrieval, we can consider clustering as an unsupervised learning problem in which we provide a structure to unlabeled and unknown data [1, 2]. Clusters formed as part of clustering contains the objects which are similar to each other in terms of their content [3, 4]. The following example as shown in Fig. 1 clearly depicts how clusters can be formed.

First, we will discuss basics of Fuzzy C Means clustering and then our approaches to modify it to get better results in terms of its accuracy. Fuzzy C Means (FCM) clustering method assigns fuzzy membership for documents belonging to clusters [3, 5]. The fuzzy membership values range between 0 and 1. Therefore, each cluster is considered as the fuzzy set of all documents. It was developed by Dunn in 1973. The Fuzzy C Means clustering method starts by assuming C as the number of clusters required, selecting random cluster centers, and assigning truth membership values to each document with respect to every cluster center. The membership values for each cluster and each document must be equal to one. In each iteration, cluster centers are updated. This algorithm iterates up to minimum objective function which can be define as [6, 7]:

N. Akhtar
Department of Computer Engineering, ZHCET, Aligarh Muslim University,
Aligarh 202002, India
e-mail: nadeemakhtar@zhcet.ac.in

M. N. Qureshi
University Polytechnic, Aligarh Muslim University, Aligarh 202002, India
e-mail: navedmohd786@gmail.com

M. V. Ahamad (✉)
Womens Polytechnic, Aligarh Muslim University, Aligarh 202002, India
e-mail: vasim.iu@gmail.com

© Springer Nature Singapore Pte Ltd. 2017
R. Ali and M. M. S. Beg (eds.), *Applications of Soft Computing for the Web*,
https://doi.org/10.1007/978-981-10-7098-3_10

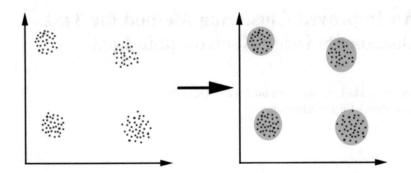

Fig. 1 Clusters of similar objects

$$J_m = \sum_{i=1}^{N} \left(\sum_{j=1}^{C} u_{ij}^m \| x_i - c_j \|^2 \right) \tag{1}$$

where $1 \leq m < \infty$ and $m > 1$, u_{ij} is the degree of membership of ith document x_i with respect to the jth cluster c_j.

The drawback with Fuzzy C Means is that if the degree of membership for a particular document for a cluster is somewhat equal for two clusters so there is ambiguity over here. So here it is difficult to tell how much it is true that document d belongs to a cluster, any cluster x. So to handle this ambiguity we need another term called as indeterminacy value which is provided by Neutrosophic logic. In case of Neutrososphic logic, we have truth, falsity, and indeterminacy values for a single document belonging to a cluster. So based on these three values we can accurately classify the document to a particular cluster, i.e., the document will belong to the cluster when it has high t, i, f value for that cluster as compared to other.

2 Background

2.1 Fuzzy Logic

Fuzzy logic [8] is the expansion of the classical and multivalued logics. It is based on the basic probability theorem that a particular event can have a probability range from 0 to 1. Fuzzy logic allows variables to have values between 0 and 1. The variable is considered to be false if its value is 0 and considered as true for value equals to 1. Fuzzy logic also considers intermediate values such as $x = 0.87$, to incorporate partial truthiness or falsity of a variable. In comparison to classical logic where we have only two outcomes either true or false, in fuzzy logic we can have various values in between to completely true and false values to deal with partial and uncertain data.

2.2 Neutrosophic Logic

The fuzzy logic is proposed to deal with the vagueness and uncertainty [8, 9]. It has two values associated with each variable, the degree of truthfulness and degree of falsehood. It can be represented as FS = {T, F}, where T and F are the degree of truthiness and degree of falsehood of the variable toward the set FS, respectively. Neutrosophic logic introduces a new parameter to a fuzzy set, called as indeterminacy. Neutrosophic logic theory considers every possible outcome for a variable X like X, Anti-X, and Neut-X which is neither X nor Anti-X [10]. According to this theory if there is indeterminacy for a particular variable or idea than that also can be expressed with a degree of membership for a variable.

3 Proposed Work

In this section, we are introducing document clustering using Neutrosophic logic. This section explores two approaches of data clustering with the help of Neutrosophic logic. The results are very promising and show the possibility of quality improvement in data clustering.

In the first approach, we added the indeterminacy factor of Neutrosophic logic to Fuzzy C Means clustering method and modified the formula which calculates the cluster centers and the truth membership of documents toward clusters. The aim of this approach is to introduce the indeterminacy factor to the Fuzzy C Means Clustering Algorithm and grouping documents (say N) in an input dataset into C clusters. The indeterminacy of documents mainly affected by the clusters which are most similar to that document. Using this concept, we calculated the indeterminacy factor using the average value of closest and second closest membership values of document with corresponding clusters. The modified algorithm tries to associate the document with the cluster having higher truth membership grade and lowest indeterminacy values toward the cluster.

As in traditional Fuzzy C Means clustering, this modified also starts with calculating the cluster centers first. Following is the modified formula for calculating the cluster center:

$$c_j = \frac{\sum_{i=1}^{n} \left(I_{ij} \cdot u_{ij} \right)^m \cdot x_i}{\sum_{i=1}^{n} \left(I_{ij} \cdot u_{ij} \right)^m} \tag{2}$$

where $[u_{ij}]$ is membership value matrix of ith document to jth cluster and $[I_{ij}]$ is indeterminacy value matrix of ith document to jth cluster and x_i is the ith document.

Further, values of membership and indeterminacy can be updated in iteration with the following modified formula:

$$u_{ij} = \frac{1}{\sum_{k=1}^{c} \left(\frac{||x_i - c_j||}{||x_i - c_k||} \right)^{\frac{2}{m-1}}} \tag{3}$$

$$I_{ij} = \frac{1}{\sum_{k=1}^{c} \left(\frac{||x_i - c_j||}{||x_i - \overline{c_{avg}}||} \right)^{\frac{2}{m-1}}} \tag{4}$$

$$\overline{c_{avg}} = \frac{c_{pi} + c_{qi}}{2} \tag{5}$$

where p_i and q_i are the clusters with the largest and the second largest membership values for document D, c is the number of clusters required, x_i is ith document, c_j is jth cluster, and m is weighted factor. In this case, we have assumed the value of m as 2.

The proposed algorithm starts by taking the input dataset having D documents and preprocessing it. As in Fuzzy C Means clustering, this algorithm also takes C (number of clusters required) random values as cluster centers. After that, membership values matrix and indeterminacy matrix is initialized. Then, it tries to associate the document with the cluster having higher truth membership grade and lowest indeterminacy values toward the cluster. This algorithm iterates and updates the cluster centers and indeterminacy values using above-mentioned equations. This algorithm repeats until objective function is optimized, which can be define as:

$$J_m = \sum_{i=1}^{n} \sum_{j=1}^{c} (u_{ij})^m \cdot (||x_i - c_j||)^2 \tag{6}$$

where u_{ij} is membership value of ith document to jth cluster, x_i is the ith document, and c_j is the jth cluster.

The second approach consists of three phases which are shown below in Fig. 2.

3.1 Preprocessing and Data Collection

The objective of this phase is to generate dataset for clustering. The format of dataset is according to standard so that if we apply our method on a preprocessed dataset that we can apply Phase II and Phase III directly. Basic steps for phase 1 are listed below:

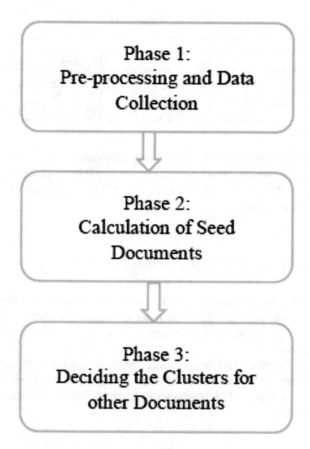

Fig. 2 Different phases of second approach of the proposed algorithm

- Collect URLs of different topics from Google search.
- Get text of URLs
- Remove images and HTML tags from text stream
- Remove helping verbs and stop words
- Perform stemming using Porter's suffix striping algorithm
- Calculate the percentage of appearance of words in a document
- Arrange words and document in dataset.

We perform a Google search using a topic string and then save the URLs of top 100 documents in the search result. We do this for all topics on which we want to generate cluster. Then we extract the data of each url as text string and then remove images and HTML tags. After that we remove helping verbs and stop words. Porter's suffix-stripping algorithm is used for performing stemming over dataset.

The words which are rooting back to the same stem can be considered as same word. For example, "compute", "computing", and "computed" can be stemmed to "comput". After the above step, find out all the words appearing in the document

Table 1 Word dataset

Word dataset	Words id	
Document id	Word id	Frequency
01	1, 2, 4, 5	20, 30, 20, 10
02	2, 4, 5, 6, 3	29, 18, 70, 12
03	5, 3, 2, 4, 14, 9	12, 13, 19, 30, 13, 78

and then calculate the percentage of frequency of words in document. Perform this step for all documents and create a dataset as a table having document id in row side and word id in column side and their respective frequency at the cell position of table as shown in Table 1.

3.2 Calculation of Seed Documents

In this phase, we are deciding the seed for the clusters. These seeds play the role of initial centroid in our algorithm. All other document's cluster is decided with Neutrosophic logic on these seed clusters. The base of these seed documents is Euclidean distance.

Euclidean distance: Euclidean distance can be calculated as the square root of differences between the coordinates of a pair of objects [3, 4]. Each object can be represented as a vector. The Euclidean distance d_{ij} can be calculated using the following formula:

$$d_{ij} = \sqrt{\sum_{k=1}^{n} \left(x_{ik} - x_{jk} \right)^2} \tag{7}$$

where x_{ik} is the kth dimension of ith document, x_{jk} is the kth dimension of jth document, and i and j are n-dimensional vectors.

Basic steps for phase 2 is listed below.

- Select a document randomly and say it as cluster 0
- For $i = 0$ to $n-1$
- Find Euclidean distance (r_i) of all other documents from the cluster i
- Select a document for which $\sum_{k=o}^{i} r_k$ is maximum and say it as cluster $i + 1$
- Consider clusters 1 to n as seeds.

The main motive of this phase is deciding the seed documents from all available documents. Initially, these documents as seed document play a role of the cluster itself. As we can visualize, the probability of being in the same cluster for documents is inversely proportional to the Euclidean distance between two documents. So we are deciding our cluster's seed documents on the basis of the Euclidean distance between documents. We are randomly selecting a document and calling it as cluster 0. Cluster 0 is a dummy cluster to process seed cluster. As shown in flow diagram, starting from this cluster 0 we find out n more documents as cluster 1 to n. These documents play initial role of clusters. As we are getting more documents in a cluster we are changing the cluster properties.

3.3 Deciding Cluster for Other Documents

After finding out seed documents that act as initial clusters, now the remaining documents have to be assigned to one of the clusters from these seed clusters. We are considering a word as a source of information for deciding the cluster for a document. Definition of some terms to be used in this phase:

- W_{ij} is the percentage of ith word in jth document.
- PC_{ik} is the average of ith word in kth cluster. (Positive Center of ith word in kth cluster).
- NC_{ik} is the average of ith word in all other than kth cluster. (Negative Center of ith word in kth cluster).
- AC_i is the average of ith word in all clusters. (Center of ith word in all clusters).
- R_i = Range of W_i. (maximum (W_i)-minimum(W_i)).

Now we have formulated truth, falsity, and indeterminacy value for a word as used in Neutrosophic logic as defined below:

Truth value for ith word in cluster kth for jth document

$$T_{ijk} = 1 - \frac{(W_{ij} - PC_{ik})}{R_i} \tag{8}$$

False value for ith word in cluster kth for jth document

$$F_{ijk} = 1 - \frac{(W_{ij} - NC_{ik})}{R_i} \tag{9}$$

Indeterminate value for ith word for jth document

$$I_{ij} = 1 - \frac{(W_{ij} - AC_i)}{R_i} \tag{10}$$

Now, these sources of information are not depended on each other so we can combine their T and I, and we can combine their T, I, and F values for a document and cluster. Let us consider the total number of words to be m. Now we are defining some terms.

Truth value for jth document to be in kth cluster is:

$$T_{jk} = 1 - \frac{\sum_{i=1}^{m}\left(\frac{(W_{ij}-PC_{ik})}{R_i}\right)}{m} \tag{11}$$

False value for jth document to be in kth cluster is:

$$F_{jk} = 1 - \frac{\sum_{i=1}^{m}\left(\frac{(W_{ij}-NC_{ik})}{R_i}\right)}{m} \tag{12}$$

Indeterminate value for jth document to decide its cluster is:

$$I_j = 1 - \frac{\sum_{i=1}^{m}\left(\frac{(W_{ij}-AC_i)}{R_i}\right)}{m} \tag{13}$$

It is clear from these formulas that if a document j is in k^{th} cluster then its corresponding T value should be high and its false value should be low. Therefore, we have introduced a new term "Deciding Factor" as DF given below:

$$\text{Deciding factor (DF)} = T - F \tag{14}$$

But as Truth values are calculated through own documents of a cluster, it should have more weight than False value. So we have modified it as

$$\text{Deciding factor (DF)} = (1.15T) - F \tag{15}$$

DF_{jk} is the "Deciding Factor" of jth document to be in kth cluster

$$DF_{jk} = (1.15T_{jk}) - F_{jk} \tag{16}$$

Algorithm for the phase 3 is given below:

- Calculate the Deciding factor DF for all documents in all clusters.
- Sort all the documents in a cluster according to their DF values for that cluster.
- Select a top 20% of document in all clusters.
- Check any document in these top 20% documents in cluster k is appearing in top 20% documents of any other cluster or not. If yes then set it as claimed (-1) otherwise set it as clear (1).
- Scan all cluster's top 20% list, starting from rank 1, if ith rank documents in all cluster are claimed then check $(i + 1)$th rank.
- Select the clusters whose ith rank document has a clear flag. Set these cluster for these document accordingly. If there is no document in top 20% of all cluster with the clear flag than from first rank documents in all clusters select a cluster whose first rank document has highest DF/I_j value and set it in respective cluster.

- Update the cluster parameters and repeat all the above steps for all other remaining documents.
- End.

Finally, after this algorithm is over, the documents are assigned to their respective clusters with higher truth value and lower indeterminacy with respect to cluster centers. In the following section below we have shown the experimental results of our methodology as compared to Fuzzy C Means clustering using Neutrosophic logic. We have also calculated and compared the accuracy in terms of precision and recall values of both of the algorithm discussed.

4 Result Evaluation

4.1 Dataset Description

We have executed the proposed algorithm on three datasets as listed in Table 2. The dataset 1 is a subset of mini newsgroup dataset available at UCI machine learning database. In this, we have 10 newsgroups having a total of 1000 evenly distributed documents to check the precision variance of the documents. Dataset 2 is also a subset of mini newsgroup dataset but in dataset 2, we put a total of 995 unevenly distributed documents in the 10 newsgroups. The dataset 3 is collected via Google search as described in Sect. 3.1.

4.2 Performance Evaluation

Precision: Peterson and Hearst gave a definition in which they defined precision as a sum of precision of relevant document viewed divide by the total number of documents, viewed, or not viewed. They treated each cluster as a category

Table 2 Description of datasets used

Dataset	#documents	Clusters	Source
Dataset 1	1000	10	https://archive.ics.uci.edu/ml/machine-learning-databases/20newsgroups-mld/
Dataset 2	995	10	https://archive.ics.uci.edu/ml/machine-learning-databases/20newsgroups-mld/
Dataset 3	1000	10	Collected via Google search using 10 different topics on which we wanted to generate cluster

dynamically generated by their method and each category in document cluster is treated as a class. The precision formula for cluster "*y*" and class "*x*" is as follows:

$$P(x, y) = \frac{N_{xy}}{N_y} \tag{17}$$

Here N_y is total number of documents in cluster "*y*" and N_{xy} is total number of common documents in cluster "*y*" and class "*x*".

Recall: Recall is defined as the sum of total number of documents common in cluster "*y*" and class "*x*", divided by the total number of documents in class "*x*". The recall formula for cluster "*y*" and class "*x*" is as follows:

$$R(x, y) = \frac{N_{xy}}{N_x} \tag{18}$$

Here N_x is total number of documents in class '*x*' and N_{xy} is total number of common documents in cluster "*y*" and class "*x*".

F-Measure: F-Measure is a quality measure, having collective impact of Precision and Recall. It combines the Precision and Recall values for each cluster with their corresponding class. The F-Measure formula for cluster "*y*" and class "*x*" is as follows:

$$F(x, y) = \frac{2 * P(x, y) * R(x, y)}{P(x, y) + R(x, y)} \tag{19}$$

Here $P(x, y)$ is precision of cluster "*y*" and class "*x*" and $R(x, y)$ is Recall of cluster "*y*" and class "*x*".

As we can see in Figs. 3, 4, and 5 above, the comparison made between Fuzzy C Means clustering algorithm and both the approaches we discussed. The accuracy of

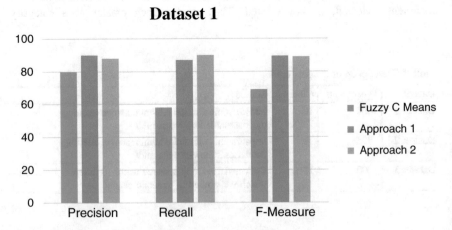

Fig. 3 Accuracy comparison of both approaches with FCM on dataset 1

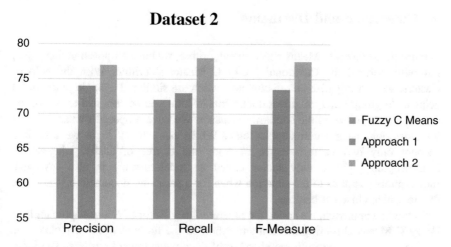

Fig. 4 Accuracy comparison of both approaches with FCM on dataset 2

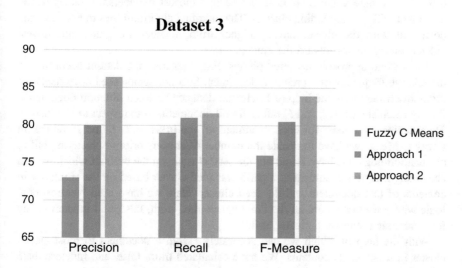

Fig. 5 Accuracy comparison of both approaches with FCM on dataset 3

both the approaches is far better than Fuzzy C Means clustering algorithm in terms of various evaluation parameters (precision, recall, and f-measure). It is clear that modified Fuzzy C Means clustering algorithm is much more accurate than the traditional Fuzzy C Means algorithm for clustering text documents.

5 Conclusion and Discussion

The improved Fuzzy C Means clustering algorithm, we have proposed in this paper, has outperformed the traditional Fuzzy C Means algorithm. With the help of clusters, we can organize text documents which are similar at a single place and it helps us to group other unknown documents in future, to be assigned to one of the known cluster based on the similarity measure. We have proposed two approaches for clustering using Neutrosophic logic. While using fuzzy logic we take into account only two values; degree of truth and degree of falsity, whereas, in Neutrosophic logic, a new factor called as indeterminacy is also involved. Indeterminacy applies to the situation when for a particular document it is not sure that to which cluster it belongs.

In the first approach, we added the indeterminacy factor of Neutrosophic logic to Fuzzy C Means clustering method and modified the formula which calculates the cluster centers and the truth membership of documents toward clusters. The indeterminacy of documents is largely affected by the document clusters that are most similar to that document. Using this concept, we calculated the indeterminacy factor using the average value of closest and second closest membership values of the document with corresponding clusters. The modified algorithm tries to associate the document with the cluster having higher truth membership grade and lowest indeterminacy values toward the cluster.

The second approach has three phases. First, generate the dataset according to the relative frequency of words in a document. Second, decide seed documents for different clusters with the help of Euclidean distance between different documents. Finally calculate the T, I, and F values for all documents with respect to all clusters. Then decide the cluster for each document on the basis of T, I, and F values. In Fuzzy C Means we have to decide the number of clusters prior to clustering but in our methodology, we have found out the seed clusters on the basis of which we can accurately assign a document to a particular seed cluster based on the similarity in contents of that document and the seed cluster. Here we have used Neutrosophic logic where we can model all the three values, i.e., truth, false, and indeterminacy for a document and for a cluster also.

With the help of which we can accurately assign a document to most closed cluster in terms of its contents. We have calculated truth, false, and indeterminate value for every document in a cluster and also for every cluster of documents with the help of which it became very accurate measure for clustering documents in one of the best possible cluster. Also, the introduction of new term, i.e., deciding factor for a document has given more weightage for a document to be in a cluster if its membership value is greatest with respect to others. Finally, the ranking of documents based on the deciding factor has helped to easily cluster them in one of the best possible clusters.

References

1. Hartigan JA (1975) Clustering algorithms. Wiley, London
2. Olson, DL, Delen D (2008) Advanced data mining techniques, 1st edn. Springer, Berlin, p 138. (February 1, 2008), ISBN 3-540-76916-1
3. Akhtar N, Ahamad MV (2015) A modified fuzzy C means clustering using neutrosophic logic. In: Proceedings of IEEE fifth international conference on communication systems and network technologies (CSNT). ISSN/ISBN 978-1-4799-1797-6/15, 10.1109/CSNT.2015.164, pp 1124–1128
4. Hartigan JA, Wong MA (1979) Algorithm AS 136: a K-means clustering algorithm. J Royal Stat Soc, Ser C 28(1):100–108. JSTOR 2346830
5. Suganya R, Shanthi R (2012) Fuzzy C-means algorithm—a review. Inter J Sci Res Publ 2(11)
6. Bezdek J (1981) Pattern recognition with fuzzy objective function algorithms. Plenum Press, New York
7. Jain AK, Dubes RC (1988) Algorithms for clustering data. Prentice Hall, Upper Saddle River, NJ. ISBN:0-13-022278-X
8. Zadeh L (1965) Fuzzy sets. Inf Control 8:338–352
9. Dunn J (1973) A fuzzy relative of the Isodata process and its use in detecting compact, well-separated clusters. J Cybern 3(3):32–57
10. Smarandache F (1998) Neutrosophy / neutrosophic probability, set, and logic. American Research Press, Rehoboth, NM
11. Bezdek J, Hathaway R (1988) Recent convergence results for the fuzzy c-means clustering algorithms. J Classif 5(2):237–247

Part VII
Soft Computing Based Web Security Applications

Fuzzy Game Theory for Web Security

Abdul Quaiyum Ansari and Koyel Datta Gupta

1 Introduction

In today's network, there has been a huge rise in the use of internet which requires strict web security requirements. With the addition of a large number of users, the amount of susceptibility in web-based applications also rises. The attackers can utilize these vulnerabilities to accomplish unauthorized admission to the web applications. The present web applications are really complex, distributed, responsive, interactive, rapidly changing, and constantly evolving [1]. The pervasive nature of web domain makes it more susceptible to malicious activities including virus attacks and security infringements. In the world, where web applications are used for important activities like banking, mailing, exchanging confidential information, security becomes a prime factor.

According to a report by [2] about 49% of the entire web applications available, are highly vulnerable and more than 13% of them can be breached without any human intervention. The report also exposes the fact that more than 80% of the websites available are viable to at least one type of attacks. Over the years various researchers have proposed techniques to counter security breaches in web applications.

The paper proposes a fuzzy game theoretic approach under consideration, that strategies adopted by defender are independent of ubiquitous attacker. A non-cooperative game is devised where the attacker tries to access and modify data assets referred as targets and the defender attempts to guard the targets.

A. Q. Ansari (✉)
Department of Electrical Engineering, Jamia Millia Islamia, New Delhi, India
e-mail: aqansari@ieee.org

K. Datta Gupta
Department of Computer Science & Engineering, Maharaja Surajmal Institute
of Technology, New Delhi, India
e-mail: koyel.dg@gmail.com

© Springer Nature Singapore Pte Ltd. 2017
R. Ali and M. M. S. Beg (eds.), *Applications of Soft Computing for the Web*,
https://doi.org/10.1007/978-981-10-7098-3_11

This paper is arranged as follows. A brief literature survey is presented in Sect. 2. Section 3 introduces the basic working principle of fuzzy-based game theory. Section 4 explains the proposed fuzzy game theory-based approach for ensuring web security. Section 5 presents simulation. Section 6 gives conclusions and future work.

2 Literature Survey

The most common attacks on a web application include Cross site scripting (XSS), Session hijacking, SQL injections, Cookie poisoning, Parameter tampering, etc. Generally, the effect of Cross site scripting is the revelation of important data like stealing of session information through cookies. Cross site scripting is the root cause behind other more intrinsic web attacks like the notorious MySpace Samy worm [3]. XSS attacks can be categorized based on the way the malicious scripts are inserted like reflected XSS, persistent XSS, content-sniffing XSS [4], etc. Several measures like Message Authentication Code can be used to shield the integrity of the Client-side state information. The paper [5] proposes several measures to preserve state integrity. The session hijacking involves intrusion in a web session to achieve unauthorized access to the ongoing web service. To track the session, identifiers should be randomly generated and transmitted over secure SSL protocol. The papers [6, 7] present "string-taint analysis", which boosts the analysis of string [8] using traces of undesirable accesses. The proposed methods mark and follow input malicious substrings and ensure that such strings are not integrated into any generated HTML pages or SQL queries. The techniques to associate unique secret tokens are proposed in papers [9–11]. Headers can also be checked to validate web requests in order to diminish CSRF attacks [12]. The paper [13] examines the assessment of effectiveness of security with respect to the barrier delay times, response force time. This procedure measures the potential risk of a network to threats, targets, and possible attacks. In some research works [14, 15], security measures like authorization, facts attribution, and data flow can be modified and access to labeled data is allowed after checking policies. In these methods, the database engine translates the authorization verification into corresponding SQL queries thereby improving the efficiency of cross-tier policy enforcement. For improving and preserving the security of web applications, the authors [16] proposes "cryptographic module validation programs" like generating passwords with the help of cryptographic module. The application of COTS components is not recommended by the paper since it might give rise to severe threats to the defense aspect of the application. The SQL injection technique inserts SQL statements into an entry field to attack data-driven applications. In [17], a method is proposed which inculcates application-specific data-flow policy at runtime for extenuating missing access control checks and script injections. The authors [18] address the problem involved in compression of semantic records for developing Semantic Web of Things. Static security checking has the benefit of having minimum or zero execution time overhead and the corresponding dynamic technique is capable of changing security policies dynamically, the authors [19] combine both to generate a better access policy. The approach checks statically the logical precision of the web application and authenticates dynamic security measures using a "known predicate", but only some access control measures can be verified. The paper [20] presents a significant study of the Web of Things literature emphasizing the security. Additionally, to accomplish secured Web of Things, the author develops architecture with smart gateways.

3 Fuzzy Game Theory

A fuzzy game is defined as a game which is unparalleled with the zero game: hence it is not greater than 0, or less than 0, nor equal to 0. As opposed to a crisp/hard game, the fuzzy game is exceptionally efficient in controlling the uncertainties [21]. The theory of fuzzy logic is applied to identify the player's preference of one payoff among the rest. The precedence of choice for every player is considered and finally, the measure of the relation between satisfaction functions is utilized to formulate crisp game.

4 Proposed Fuzzy Game Theory for Web Security

To develop an intricate security management solution for web applications, a fuzzy game theory approach is proposed. The system includes evaluation of the security level of the system, and unauthorized access and update detection and prevention. We define the web security issue as a fuzzy game theoretic model that addresses the decision-making to prevent unauthorized data access and update and defend the website system. This is a two-player, non-cooperative static game between a system defender P_1, which defends unauthorized access to the website against a ubiquitous attacker P_2 and the set of the available targets is represented by $T = t_1, t_2, \ldots, t_n$.

The player P_2 may select to attack any target from T whereas P_1 will try to resist the attacks by defending targets using system administrator facilities. In the proposed game, the targets are the information accessible through a web application. A set of strategies for P_1 is defined as S_1 is the exclusive schedule where the system administrator resource will revise the targets' security level from lowest to highest. The utility of P_1 when t_i is attacked and has no resources assigned to it then the payoff equals to the lowest security level $llP_1(t_i)$. On the other hand, if P_1 is able to protect t_i by some available system administrator, the payoff equals to $hlP_1(t_i)$. Finally, the payoff in case where no attack has taken place equals zero. Likewise, for P_2, we define their payoff values as $llP_2(t_i)$ and $hlP_2(t_i)$, respectively. The fuzzy adaptation of this game is defined as for each player P_1, the fuzzy constraint set perceived by them is denoted by $\mu_1: S_1 \rightarrow [0, 1]$. The constraint μ [22] differ in their degree of possibility and only some are practically feasible with μ equal to one. This regulates the choice of strategies. The number of strategies for n resources and of k targets is given by $C(k, n)$. The constraints considered in this game include unavailability of port to secure network traffic, number of resources/system administrators available. The strategies for P_2 are denoted by S_2 and p_a represents the probability of attacking a target.

Moreover, the difference between payoffs for higher level security and lower level security w.r.t. t_i for both players are denoted as Δ_1 and Δ_2.

4.1 Defining Payoff in the Fuzzy Game

When a strategy profile $\langle s_1^*, s_2^* \rangle$ is played, the payoff values of both players P_1 and P_2 are given as:

$$\delta_{P1}\left(s_1^*, s_2^*\right) = \Sigma p_a(\mu_1 \text{hlP}_1(t_i) + (1 - \mu_1 \text{hlP}_1(t_i))) \tag{1}$$

$$\delta_{P2}\left(s_1^*, s_2^*\right) = \Sigma p_a(\mu_1 \text{hlP}_2(t_i) + (1 - \mu_1 \text{hlP}_2(t_i))) \tag{2}$$

The maximum payoff a player is given by

$$\text{Payoff PO} = \max_{s_i \in S_1} \min_{s_j \in S_2} \delta_i(s)$$

4.1.1 Case Study

Let us assume there are three resources with the security guard, i.e., P_1 and the number of targets available is four, then the number of strategies are $C(4, 3) = 4$. Let us assume the following fuzzified payoff matrix in Table 1. Each of the players has four strategies.

Now by using minimum priority of all payoffs in a cell, the following matrix (Table 2) is computed to identify the maximum payoff a player can ensure for himself.

Now, by taking the maxima of Row minima for P_1

Table 1 Fuzzy payoff matrix

Strategy	1	2	3	4
1	(0.6, 0.24)	(0.3, 0.3)	(0.22, 0.24)	(0.24, 0.24)
2	(0.06, 0.06)	(0.22, 0.3)	(0.08, 0.24)	(0.22, 0.3)
3	(0.32, 0.06)	(0.3, 0.26)	(0.24, 0.06)	(0.22, 0.22)
4	(0.24, 0.24)	(0.3, 0.3)	(0.06, 0.06)	(0.3, 0.24)

Table 2 Minimum priority payoff matrix

	P$_2$ chooses j					
P$_1$ chooses i	Strategy	1	2	3	4	Row min
	1	0.24	0.3	0.22	0.24	0.22
	2	0.06	0.22	0.08	0.22	0.06
	3	0.06	0.26	0.06	0.22	0.06
	4	0.24	0.3	0.06	0.24	0.06
	Column max	0.24	0.3	0.22	0.24	

$\max_i \min_j = \max\{0.22, 0.06, 0.06, 0.06\} = 0.22$, i.e., P$_1$ is guaranteed not to get an amount less than 0.22 by choosing strategy 1.

For P$_2$ minima of Column maxima is generated.

$\min_j \max_i = \min\{0.24, 0.3, 0.22, 0.24\} = 0.22$, i.e., P$_2$ cannot get an amount less than 0.22 by choosing strategy 3.

5 Simulation

To measure the performance of the proposed method a java-based environment is created with 1000 sample of attacks. The samples are iteratively used 20 times and the average damage to the targets (data source) is measured and plotted (Fig. 1) where the number of targets has been varied as 5, 8, 10, and number of resources available to defend the targets is considered as 1, 2, and 3. The expected

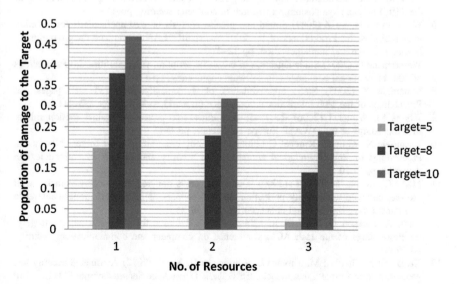

Fig. 1 Proportion of damage using the proposed technique

unauthorized modifications of data sets are calculated as the result of an attack on a given target following the strategies used by the defender.

6 Conclusions

The proposed algorithm with the fuzzy game theory approach is able to tackle any random unauthorized data access to a large extent. Moreover, the decision-making process has been simplified by using the concept that only viable strategies will be used by the defender. For simplicity, it has been assumed that an attacker attacks a single target for any particular strategy; however, it can be extended for attacking multiple targets using a single strategy. In future other web security issues can be considered for implementation.

References

1. Marchetto (2008) Special section on testing and security of web systems. Int J Softw Tools Technol Transfer 10(6):473–476
2. WhiteHat Security Web Applications Security Statistics Report (2016) https://info.whitehatsec.com/rs/675-YBI-674/.../WH-2016-Stats-Report-FINAL.pdf
3. MySpace Samy Worm (2005) https://samy.pl/popular/tech.html
4. Barth A, Caballero J, Song D (2009) Secure content sniffing for web browsers, or how to stop papers from reviewing themselves. In: Proceedings of the 30th IEEE symposium on security and privacy, pp 360–371
5. Johns M (2006) SessionSafe: implementing XSS immune session handling. In: Proceeding of the 11th European symposium on research in computer security, pp 444–460
6. Wassermann G, Su Z (2007) Sound and precise analysis of web applications for injection vulnerabilities. In: Proceedings of the 2007 ACM SIGPLAN conference on programming language design and implementation, pp 32–41
7. Wassermann G (2008) Static detection of cross-site scripting vulnerabilities. In: Proceedings of 30th international conference on software engineering, pp 171–180
8. Minamide Y (2005) Static approximation of dynamically generated web pages. In: Proceedings of the 14th international conference on world wide web, pp 432–441
9. Johns M, Winter J (2006) RequestRodeo: client-side protection against session riding. In: Proceedings of the OWASP Europe conference, pp 1–15
10. Jovanovic N, Kirda E, Kruegel C (2006) Preventing cross site request forgery attacks. In: Proceedings of 2nd international conference on security and privacy in communication networks, pp 1–10
11. Mao Z, Li N, Molloy I (2009) Defeating cross-site request forgery attacks with browser-enforced authenticity protection. In: Proceedings of 13th international conference on financial cryptography and data security, pp 238–255
12. Barth A, Jackson C, Mitchell JC (2008) Robust defenses for cross-site request forgery. In: Proceedings of the 15th ACM conference on computer and communications security, pp 75–88
13. Xu D, Tu M, Sanford M, Thomas L, Woodraska D, Xu W (2012) Automated security test generation with formal threat models. IEEE Trans Dependable Secure Comput 9(4):526–540

14. Swamy N, Corcoran BJ, Hicks M (2008) FABLE: a language for enforcing user-defined security policies. In: Proceedings of 29th IEEE symposium on security and privacy, pp 369–383
15. Corcoran BJ, Swamy N, Hicks M (2009) Cross-tier, label-based security enforcement for web applications. In: Proceedings of 35th SIGMOD international conference on management of data, pp 269–282
16. Dima A, Wack J, Wakid S (1999) Raising the bar on software testing. IT Professional 1(3): 27–32
17. Yip A, Wang X, Zeldovich N, Kaashoek MF (2009) Improving application security with data flow assertions. In: Proceedings of the ACM SIGOPS 22nd symposium on operating systems principles, pp 291–304
18. Scioscia F, Ruta M, (2009) Building a semantic web of things: issues and perspectives in information compression. In: Proceedings of IEEE international conference on semantic computing, pp 589–594
19. Chlipala A (2010) Static checking of dynamically-varying security policies in database-backed applications. In: Proceedings of 9th USENIX conference on operating systems design and implementation, pp 105–118
20. Xie W, Tang Y, Chen S, Zhang Y, Gao Y (2016) Security of web of things: a survey. In: Proceedings of international workshop on security advances in information and computer security, pp 61–70
21. Chakeri A, Dariani AN, Lucas C (2008) How can fuzzy logic determine game equilibriums better? In: (IS'08), 4th international IEEE conference on intelligent systems, pp 100–105
22. Aristidou M, Sarangi S (2006) Games in fuzzy environments. South Econ J 72:645–659

Part VIII
Soft Computing Based Online Market Intelligence

Fuzzy Models and Business Intelligence in Web-Based Applications

Shah Imran Alam, Syed Imtiyaz Hassan and Moin Uddin

1 Introduction

Vagueness is not an ingredient of exact reasoning, but most of the real world scenarios are complex enough to be precise. Imprecise scenarios define and dictate many of the problem domains, and this drives the need for business intelligence to be intelligent enough to equip decision-making process with this dimension. While we researchers try to model parts of the software applications with human-like intellect to deal with such set of problems. While they attempt to automate the process of finding reliable and cost-effective solution, it becomes necessary to allow the inputs of such processing units to be imprecise and the processing unit to be capable of dealing with those inaccurate figures. In next two subsections, we discuss two mathematical approaches namely, rough set theory and fuzzy set theory, both of which deal with imprecise input sets and software components with vague input-set processing logic.

1.1 Rough Set Theory

Rough set theory has been proposed as an approximation of the crisp set theory. It is defined in terms of lower approximation and upper approximation [1].

S. I. Alam (✉) · S. I. Hassan · M. Uddin
Department of Computer Science & Engineering, School of Engineering Sciences &
Technology, Jamia Hamdard (Deemed to be University), New Delhi, India
e-mail: shahimranalam@gmail.com

S. I. Hassan
e-mail: s.imtiyaz@gmail.com

M. Uddin
e-mail: Prof.moinuddin@jamiahamdard.ac.in

© Springer Nature Singapore Pte Ltd. 2017
R. Ali and M. M. S. Beg (eds.), *Applications of Soft Computing for the Web*,
https://doi.org/10.1007/978-981-10-7098-3_12

Table 1 Patient data decision table

Patient ID	Medicine 1 (mg)	Medicine 2 (mg)	Medicine 3 (mcg)	Recovery
X1	5	2	100	Yes
X2	10	5	500	No
X3	5	15	500	No
...				
Xn	25	5	200	Yes

The original draft of the definition proposed both the lower approximation and the upper approximation sets to be crisp set but later variations extended the definition of the two approximations to be fuzzy sets. Instead of approximation, membership function could also be used to formulate the degree of imprecision [2].

This theory soon becomes a handy tool to solve the problems of reducing the data, design efficient rules, and to even leverage discovery of valuable patterns in vague data set. Rough sets are classified in terms of objects and attributes. For example, in a patient's data, the list of patients could be the object as shown in the first column and the attributes being the medications prescribed as shown in the subsequent columns in the decision Table 1. This decision table captures the data for the patients with common illness or symptoms.

The above decision table has multiple attributes with each attribute showing the dosage of the particular medicine. These attributes are represented in crisp figures and could be best described as conditional class whereas the last column represents the attribute that indicates the "recovery". As shown in the table, the "recovery" attribute typically represents the decision class and hence is a better candidate to be modelled as a rough set. The boundary region might not be an empty set or even a crisp set with finite elements say $S = \{Yes, No\}$ as shown in the table. The above-discussed example also serves the purpose to identify the difference between the part of the data that should be modelled as crisp and part of the data that makes more sense as non-crisp data while building the human-like decision-making capability in the software system.

1.2 Fuzzy Set Theory

Imprecision is the need of the hour and of the modern day complex applications, in both the cases of enterprise application and non-enterprise application. In this era of communication, the majority of such applications are designed to work as a web application to reach distributed teams and larger user groups across the globe. Lofti Zadeh had introduced a solution to such vague modelling requirement with fuzzy set theory [3], which came as an extension to the trivial crisp set theory. Its further extensions detail out the fuzzy set theory as a theory capable of dealing with possibilities, similar to the way in which a random variable is aligned to the probability distribution [4]. This technique could be well applied to project the predicted behavior over time with the similar past business decision records and their achieved results. In the fuzzy set theory, the elements are identified with the membership functions often called as characteristics function that assigns a degree of belong-ingness to the element of the set, which makes the membership value non-crisp.

In notations, the fuzzy set S is denoted in terms of membership function $S = \{x\mu_S(x)\}$ for the element x. The basic set operations in fuzzy sets are also defined in consideration to the membership value. The union operation of two fuzzy set is defined by the MAX function that assigns the larger value of both the set elements, as specified by the membership function [5].

That is the union set,

$$S_3 = \{\mathrm{MAX}(\mu_{S1}(x), \mu_{S2}(x))\} \tag{1}$$

Assuming x, being one of the goals of multiple solutions, the union operation would result in an optimistic approach while deriving the solution. Similarly, the intersection operation of two fuzzy set is defined by the MIN function that gives the smaller value of both the set elements, as defined by the membership function [5].

That is the intersection set,

$$S_3 = \{\mathrm{Min}(\mu_{S1}(x), \mu_{S2}(x))\} \tag{2}$$

The intersection operation would result in a pessimistic approach while deriving the solution. The graph below describes both fuzzy union and fuzzy-intersection operations with visual clarity and simplified version of Mamdani Fuzzy model which presents these two operations in terms of MIN and MAX functions as in Eqs. 1 and 2 above (Mamdani Fuzzy Model) (Fig. 1).

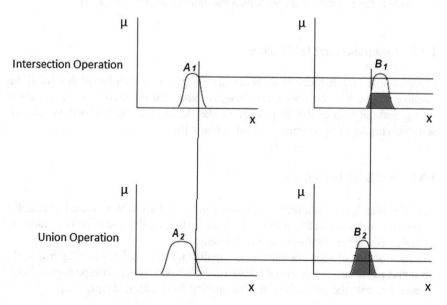

Fig. 1 Fuzzy intersection and union operations (Mamdani Fuzzy Model)

1.3 Related Terms

This section discusses a few terms which are of relevance to overall discussion related to intelligence in business applications.

1.3.1 Competitive Intelligence

It is derived from market behavior analysis and focuses on the business domain understanding. That means vital past and survey data are more critical to build upon competitive intelligence system to be on the top of its field [6].

1.3.2 Business Intelligence

Business intelligence is complementary to competitive intelligence in the sense that it is more about the intelligence that is built in software cores which are responsible for delivering edge in decision-making and planning. The inputs to such components are competitive intelligence data along with the expert inputs. In fact, Business intelligence, when seen as a completely independent system may function beyond the decision-making process and hence this extension is more suitably described as a workflow. Such system extends the notion of business intelligence to "intelligent business intelligence" by combining both advanced level BI technique and workflow technique to beat the performance need of a complex decision-making system with least possible human interaction [7, 8].

1.3.3 Computational Intelligence

Computational intelligence is all about the design of the "artifacts" that build the intelligent agents. Such agents are designed based on the technique of soft computing and may form the core logic of the decision system, expert system, or self-evolving learning systems and subsystems [9].

1.3.4 Artificial Intelligence

Artificial intelligence is a more generalized field of study which is used to produce human-like behavior in the systems. In fact, the human-like behavior is an induced behavior and hence synthetic in the true sense.

The relation between computational intelligence and artificial intelligence could be stated as former being the offspring of the latter. Usually, computational intelligence harvests the strength of soft computing for building agents [10].

2 Components of a Business Intelligence (BI) Workflow

2.1 Rule Base Engine

Rule base engine is also termed as business rule engine or inference engine which is a pluggable software system that could be integrated as a key but an independent component to the BI workflow system while automating the decision and triggering a course of actions. The action is defined as the routine that executes in accordance with the externally defined business rules. Business rules are externally defined or more appropriately said are "declarative" in the sense that they are not hard-coded within the system. They are coded as a part of the functional configuration of the database or in an external file system, typically in a file written in the extended markup language (xml) or plain old Java objects (POJO). POJO is a term used in the context of Java technology [11]. Other object-oriented languages have a similar naming convention like POJO to refer to the simplicity of the objects in containing information with, no special needs or coupling with any particular frameworks. Such externalization of the business rule makes the system robust to ever-changing business needs. This de-coupling of the processing code with the rule statements allows the development of the otherwise highly complex solution to be much simpler and maintainable at low cost.

The functional configuration of a quality-focused rule base engine allows not only to declaratively frame rules but also allows specifying the levels of priority to them if desired. Moreover, any required preconditions and exceptional cases of mutual exclusions could also be specified easily. Supporting external routines and functionalities are also provisioned in these systems. The engine by itself is capable of finding inconsistencies in the given rules.

A simplified example of a declarative rule-fact-action specification is shown as below

Rules:
R1: IF Hot AND Smoke_Like THEN Fire
R2: IF Alarm_On THEN Smoke_Like
R3: IF Temp_Reading_High THEN Hot
R4: IF Fire THEN Switch_On_Fireprotection
Facts:
F1: Hot
F2: Smoke_Like
Goal:
Auto switch on fire protection system to spray water.

2.2   Fuzzy Inference Engine: A Fuzzy Extension to the Rule-Based Engine

Fuzzy inference engine binds together the inference engine with the knowledge base that defines the membership functions of the data set to the fuzzification and de-fuzzification routines to reap the benefits of the fuzzy operations in the decision-making process for the vague input set [12].

The transformation of data set and fuzzy operation distinguishes fuzzy inference engine from non-fuzzy inference engine, which works on crisp data [13]. Figure 2 shows the control flow along with input and output transformation [14]. Various variations of fuzzy inference systems have been implemented in computer science literature [15].

2.3   Data Mining Engine

Data mining engines are the automated and pluggable third-party systems that are used to discover interesting patterns [16]. Such engines normally come as a subsystem of machine learning libraries that could even be integrated with an external component like MapReduce of Apache Hadoop for mining big data [17].

Data mining engines are also available as workflow-based programs that could efficiently execute plugged Weka code and functional routines written in R or Python [18]. However, some application's business domain may require specialized frameworks as per the nature of data and the needs of domain and may not be suitably modelled as a pluggable and reusable engine. As an example, we refer this mining framework for criminal activity pattern discovery which is based on Coplink project [19]. Data mining system in general leverages on the strength of

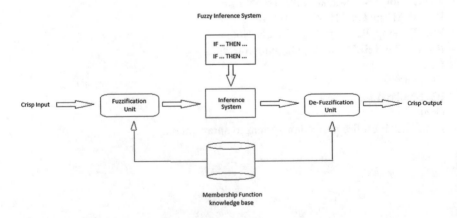

Fig. 2 Fuzzy inference system with dependent units [13]

fuzzy logic. One such system which prefers fuzzy data over crisp data and which designed to detect the fault in the generation of digital protection consideration to build its logic gives results that is much more sensitive in finding ground faults as compared to crisp base mathematical model [20].

2.4 Business Analytics System

Business analytics system works as an explorative agent. This system emphasizes on a statistical representation of the organization's data to make business decisions [21]. However, it is often built as a superset to the data mining engines in the sense that it may deliver an end to end solution for the whole of "decision support requirement" through the analysis of data warehouse records and by extending data mining support to it. Normally, such massive systems are propriety. Open source and free solutions are also radially available as interdependent libraries with each one being more focused on solving a specialized set of problems or domains without the end to end solutions, under one umbrella.

2.5 Expert System

The expert system can be defined as an emulator of a decision-making capability, like of a human expert, powered by captured knowledge in an expert database over a period.

Expert systems have always been a key research area of artificial intelligence and have its base on the techniques of soft computing.

Fuzzy expert systems are proposed for the various areas of applied engineering in many domains that are of significant impact. Recent research trend shows a number of researchers suggesting its applicability in decision-making processes of medical science for diagnosis of a number of critical illnesses. Such work includes liver diseases [22], thyroid conditions [23], mental health issues [24], and musculoskeletal disorders [25].

2.6 Transformation Unit

Transformation unit is a routine specific to the domain of the system or subsystem that can translate inputs defined in a crisp set to non-crisp inputs and vice-versa. This conversion of the crisp set to non-crisp set is termed as Fuzzification and the process of converting back the non-crisp set to a crisp set, which is usually an output of the functional routine, is termed as De-fuzzification [13].

The transformation unit could be the part of subsystems like a business analytical system, data mining engine, expert system or BI workflow systems in a broader sense.

3 Factors Effecting Decision-Making as Non-crisp Model

Several business domain factors come into pictures before the flow enters the decision-making routine requiring additional weights to be applied to the input and output in decision-making system. It may lead to the conversion of crisp inputs and outputs to non-crisp inputs and outputs. The output referred here, also include the subsystem output that is further passed to another component, of a much larger decision-making system. This output may not necessarily be the final result that dictates the course of action. The weighted rough set could be one of the right techniques to introduce the impact of such factors to the input, and final decision may be based on aggregated weighted strength factor which is more suitable to minimize imbalance [26].

Three categories of business impact factors are classified here

1. Environmental factors: factors affecting the business due to the surrounding particulars, like the cost of labor in a particular region where the company is running. Generally, changes in such environmental factors are slow. Sharp variations are possible with events like amendments to US jobs outsourcing law that barred US IT firms to outsource a large chunk of jobs.
2. Implicit factor: factors internal to the business unit, functioning, and team. For example, a decline of purchase pattern on a shopping portal due to an incompetent logistic support. Such factors impact business drastically within less time-period.
3. Explicit factors: Factors that are not internal to business or environment but act from outside to impact different sets of businesses, together. Like war, political turmoil, terror activities, or democratically induced events, for example, Brexit.

4 Business Intelligence (BI) Applicability in Three Web-Based Enterprise Applications

In this section, we discuss three examples of web-based enterprise applications and for each example, we present a different aspect of BI workflow system. In the first example that is in CRM application, the core modules of the BI architecture is depicted in terms of a block diagram with BI being the overall workflow system. The high-level design is meant to identify the components which typically operate on fuzzy data set and to understand the flow of control. Moving the fuzzy operation along with the underlying algorithm to a centralized unit is a proposed design. This design could effectively be realized by leading cloud services providers as they could provide it as "software as a service" product that could attract not only high revenue to them but also offer more maintainable and affordable high-end cloud-based solution to clients worldwide.

In the second example of an e-commerce application, the key performance indicator (KPI) and its classification are focused with relevance to the e-commerce industry that will elaborate the necessity of the requirement of handling the imprecision of the input set.

In the third example of online trading application, three well-known algorithms-based fuzzy model would be discussed regarding their applicability to the business domain. We will extend this discussion to emphasize the need of preprocessing of the data set depending upon the nature of requirement and its objective. Moreover, the algorithms would also be discussed in the context of their limitations in making a better decision in the specific business attribute, criteria and goal data set which has the occurrence of outlier data. We also propose suitable and general algorithmic extensions to deal with the issue.

4.1 CRM Applications

Customer relation management software is enterprise solution to track the client behavior and resolve their problems, generally with the help of a ticket-based system [27]. Part of the CRM functionality is to allocate resources, products, or services on the request of the customer along with payment processing and billing system under the same umbrella.

The general classification of a CRM application is done in two broad categories that are, operations and analysis.

The block diagram below proposes a BI workflow framework designed as a pluggable solution to the CRM solution application set. The sequence of flow control is labeled [27].

The gray colored blocks in the above figure refer to those units which need profound vague data set processing capabilities in general [12, 28–30].

The above references are mere indications of components shown in gray in Fig. 3. There are numerous researches which report the applicability of fuzzy set theory in these modules. In fact, the rest of the application modules also get benefited from fuzzy set theory more or less. A further proposed extension to the system shown in Fig. 4, suggest an abstraction of "fuzzy set operations" and most of the recognized "algorithms that work on fuzzy set" as an independent and decoupled system which could be implemented to be used as a framework over the cloud. Not only would this place all changes to be incorporated in one node resulting in less effort wasted in duplication but also would allow a larger group of developers of different domain utilize the out of the box capabilities with better economy. Such a system could be termed as a "fuzzy operations and service center"

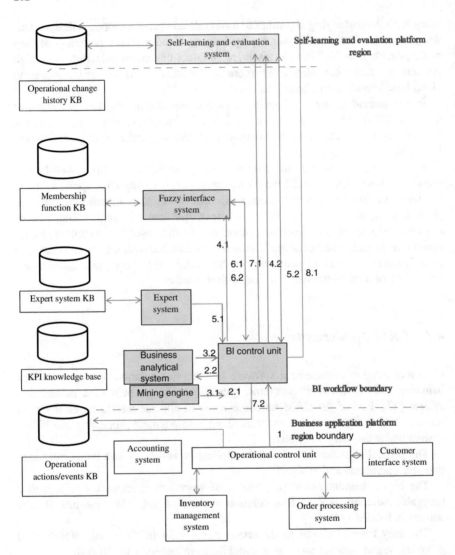

Fig. 3 Components of CRM application in layered design

in context to the rest of the component in the design. The cloud service providers, as they already own huge existing infrastructure, and capability to provide "software as a service" [31]. Figure 4 illustrates the idea.

Apart from the benefit of a nodal point of change, it would also lead to less costly and high-quality services regarding security patches, availability, and low up-front cost. The centralized system would also be decoupled from the functionality of the more specialized system as shown in above figure. The services in the Fuzzy operations and service center could be provided as web services,

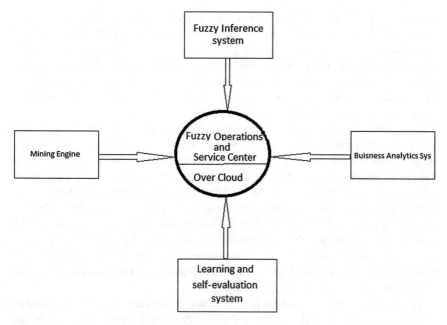

Fig. 4 Fuzzy operation and service abstraction

which allows Simple Object Access Protocol which is more secure but suited for invocation of services with large data set. For simple operation with less security need Representational State Transfer style of communication could be used which is many folds faster. The whole CRM system as in Fig. 3 has been shown as three interdependent layers with separation of concerns and could be distributed over an inter-cloud to reap the benefits of multiple service providers of specific infrastructure needs as well. In such design deployed over inter-cloud, an optimized deployment approach could be chosen to further improve the performance of the whole system as well [32].

4.2 e-Commerce Applications

e-commerce is a web-based application platform (a shopping portal) that facilitates the online sales and purchase of goods and services. In very less time, it has become a preferred platform that provides online selling and purchase services to sellers and customers, without limiting the regional boundaries. To succeed, an e-commerce platform solution needs to be equipped with cutting-edge analytical and data mining abilities. The BI system that spans across the whole workflow is very much dependent on the inputs of a number of subcomponents, business analytical system and data mining engine being the two important input generators for the decision-making cycle of BI. The input from web analytics system, data mining system and the expert system could be crisp inputs as well as interpretation could be non-crisp. It may be the need to explicitly convert crisp input set to non-crisp input set. Also, it is quite usual that the business analytical system and mining routines itself undergo fuzzification of its working data set to incorporate the vagueness of the boundary and the results may undergo a de-fuzzification to get converted to precise figures before the reports are generated.

The e-commerce application itself have CRM and ERP components to support its end to end functionality, apart from the regular need for domain-specific and custom build support system. For an e-commerce application, a typical business need is to analyze the buying pattern and customers preferences along with overall sales indicators to evolve as a competitive shopping platform. Concerning the user-friendly ability, such a portal must allow the customers to filter the products as per their quality preferences and varied choices. The feedback summaries of the previous buyers and positive automated suggestion may help the customer to be more involved. The feedbacks and the recommendations may not necessarily be modelled with precise boundaries rather non-precise classification could be more efficient in designing these parts of the system.

In the context of the business health, which is the actual base of e-commerce platform, it is meaningful to explore the in-depth details in of the nature of key performance indicators while we limit our discussion to a business analytical system and the vague and precise parts of key performance indicators. The KPI are the prime factors on which a business analytical system generates the performance report of the application [33]. Normally these key performance indicators belong to the business domain in particular but the standard business analytical system, which is plugged in as a module to the BI workflow provide provisions to capture most of the relevant details from a more general range of indicators. The reports generated by business analytical systems could be used to visually analyze the possible indirect impact on the business health and also could be fed to the BI workflow which may pass it to the rule base engine to incorporate the right course of action based on predefined business rule leading to corrective measures.

Google analytics is the example of one such business analytical system that could be integrated as a framework to the BI systems for its analysis needs.

Typically, as in most of the complex enterprise businesses, the success not only depends on the performance indicator analysis (which in turn depends on the business software platform) but also depends on the large extent to the inputs from the third-party system. These third-party systems are external systems which belong to business partner group and may function in different domains. For example, an e-commerce platform may suggest potentially preferred products or services based on the browsing or blogging activities by the same user on social media. For such functional needs, social media platform itself provides their high-end plug in the analytical framework that could be integrated with the e-commerce platform to collect KPI's from social media. Facebook analytics is one such easy to integrate framework.

Table 2 e-Commerce platform KPI

KPI's for e-commerce platform	KPI's for sales performance indicators
Online visit traffic	Time-bound sales figures for individual products
Transaction traffic	Regional clustering of preferred products or services
Logged in period	Comparison with past sales figures
Product or services browsing history for each customer	Customer reach out for the time bound of sales durations
Refer to friend history	Product purchase affinity
Product and seller rating history	Product view or wish list addition
Subjective blogging and commenting on external platform (social media)	Cost-based slab categorization of the sold products
Number of social media connects	
Preferred payment modes	
Issues resolution satisfaction ratio	
Issues resolution time and number of messages exchanged	
Frequency of visit of existing customers	

Below we suggest two sets of KPI's, one is specific to the e-commerce platform itself along with the KPI's collected from social media which we call it as e-commerce platform KPI's. The other set of KPI's is actually sales performance indicators which give the insight of the actual sales figures and the revenue generated.

As we can see from the two KPI's list below, the e-commerce platform KPI's are indirect factors while they affect the actual revenue figures whereas the sales performance indicators are directly related to the revenue generated. Hence, it could be justified to say that e-commerce platform KPIs fall into the category of attributes whereas sale performance indicators are the set of goals that need to be maximized (Table 2).

In the next section, we will see the example of how the two set of inputs, similarly categorized by attributes and goals are used in implement decision logic using various well-established models of decision theory [34].

4.3 Automated Trading Application

An automated trading system or agent that may constitute the part of a larger system is a software system that could manage client's portfolio by automatically creating buy or sell orders and submitting it to the exchange [35]. The routines in an automated trading agent are intelligent enough to decide the necessary actions in connection with the fluctuations of the market with the goal to optimize the client's investment objectives. The overall purpose of such a trading agent is to utilize artificially induced intelligence to make necessary decisions that could minimize the loss and maximize the gain. For example, in trading terminology, it may be used to decide when to go Bearish or Bullish and on which stock, depending on the market fluctuation, embedded expert intelligence, and by using rich knowledge-database.

Fuzzy-based model of an automated trading agent is a natural way to analyze empirical results in making decisions and it performs better than non-fuzzy portfolio allocation [35]. In such system, the fuzzy reference model helps utilize the fuzzy inference technique to large extent. For a better understanding, we give an example of a membership function in a fuzzy reference model which can gain from switching membership function that switches from sigmoidal to trapezoidal and could be used to compare projected outcome for maximum gain. As suggested by the improvement analysis in case of a trading agent in this work [35].

4.4 Decision-Making with Fuzzy Data Set: An Overview of the Related Work

Decision-making becomes quite complicated operation with many depending factors that must be considered while taking a decision. It further gets complicated because of vagueness in fact. Considering the data set as a fuzzy set is the natural way in the decision-making process. Because it affects the economy, business, healthcare, engineering, and almost all important spheres of life, it has many research contributions towards designing the algorithms that could possibly improve the decision-making operation while we entrust machine for it.

We present here three well known and related algorithms that are designed for automating decision-making routines.

1. Max–Min method
2. Minimization of regret method (MMR) and
3. Generalization of MMR with the introduction of Ordered Weighted Average (OWA) operator.

The max–min algorithm forms the baseline of minimization of regret method which was later on extended as a generalization of minimization of regret with the introduction of Ordered Weighted Average (OWA) by Yager [36].

4.4.1 Max–Min Method

In this section, we have discussed a fuzzy model application on a trading scenario using max–min approach to understand the applicability in the financial trading problem.

The applicability of this model has been reported by many researchers in solving problems of various domains, and its importance has been well recognized in recent research trends as well [37–39].

The model considers two base sets, one being the set of alternatives, $A = \{a,...,a_n)$, for $i = 1$ to n and the other is the set of goals, $G = \{g_1, g_2,...,g_m\}$ for $j = 1$ to m. Both alternative and goals are fuzzy sets defined by their respective membership functions. All individual goals have a weight, which represents its importance factor with respect to other goal elements.

The decision is taken by considering the goals in a pessimistic way by applying intersection on fuzzy goals thus the formulation for goal decision say D is for an alternative is

$$D = G_1^{W1} \cap G_2^{W2} \cap \cdots \cap G_m^{Wm} \tag{3}$$

And the chosen alternative must be with maximum degree or membership in decisions set D. This is realized by union operation on fuzzy set D. Consider trading scenarios with alternative security picks of derivatives including future, swaps and options and are considered as attributes.

With the final objective being the maximization of profit, the goals are defined as below.

The goals are:

$G1$: {The percentage gain in last quarter}
$G2$: {The average cost fluctuation ratio in last four quarter}
$G3$: {The overall market reputation of the lender}.

The above goal is a typical example of projected goals based on past performance of each security being a financial product (Table 3).

It could be noted that certain logical preprocessing may be needed at one or more level in the particular decision support routine to make the data suitable for the algorithm. For example, the average cost fluctuation ratio could be both, a positive value as well as a negative value. The assumption is made here is that only positive value is picked up by the BI system before invocation of this algorithm as we are determined to get the best return on our investment.

Table 3 Trading decision table

Attributes	G1	G2	G3
Future	0.3	0.7	0.8
Option	0.5	0.7	0.4
Swaps	0.9	0.2	0.2

Another preprocessing requirement that makes much sense in the case where a cost fluctuation is one of the weighted goals along with the other goals of the system. We will be required to apply fuzzy-compliment operator on the goal set before using the data for decision-making. The "compliment" operator will convert the goal to cost consistency membership values which would be required for the final objective, which is to gain maximum profit. The cost fluctuations goal would have been good consideration provided the final objective would have been to minimize the risk.

Applying Max–Min model we get

$$\mu_D(X1) = \text{MIN}\{0.3, 0.7, 0.8\} = 0.3$$
$$\mu_D(X2) = \text{MIN}\{0.5, 0.7, 0.4\} = 0.4$$
$$\mu_D(X3) = \text{MIN}\{0.9, 0.2, 0.2\} = 0.2$$

Hence

$$\mu_D(X_{\text{Optimal}}) = \text{MAX}(0.3, 0.4, 0.2\} = 0.4$$

Hence the decision $D = \{\text{Option}\}$ as per this algorithm.

4.4.2 Decision-Making Using Minimization of Regret Method (MMR)

The idea of two base sets, the attribute, and the goal set remains the same as discussed in the max–min approach in above section. However, while we illustrate the algorithm with an example in subsequent sections we preferred to change to a more detailed example taken from the e-commerce domain.

In MMR method the matrix that is formed with the alternatives and the set of goals (also termed as a state of nature) is called as the "payoff matrix" [36]. The MMR method employs additional decision-parameter called as regret matrix, where each data elements r_{ij} maps to an element c_{ij} in the payoff matrix. The regret factor r_{ij} is used to quantify the deviation from the expected result [39].

The MMR method has the following operation step.

Step 1 Populate the payoff matrix with the attributes A and goal G elements.
Step 2 Apply fuzzy union, i.e., Max operation on each column of the payoff matrix to obtain row A_{max} for each column as formulated below.

$$A_{\text{max}} = \text{Max}\{A_{ij}\} \text{ for } i = 1 \text{ to } n$$

Step 3 Calculate the regret matrix for the attributes and goals data sets such that in that payoff matrix A, r_{ij} of regret matrix R is $A_{\text{max}} - A_i$
Step 4 Perform fuzzy union operation on each row of the regret matrix to obtain column R_{max} such that $R_{\text{max}} = \text{Max}\{R_{ij}\}$ for $j = 1$ to m
Step 5 Perform intersection operation on the column, R_{max} such that Decision $D = \text{Min}\{R_{\text{max}}\}$.

That is we perform Min on Max of R_{ij} to obtain the decision D.

4.4.3   Generalization of Minimization of Regret Method (MMR) with OWA Operators

Yager's approach generalized the MMR method based on regret matrix by incorporating parameterized family of aggregation operators. The first three steps in this method remain the same as MMR method discussed in previous section [34].

Step 4 of the MMR algorithm that prefers a Min (intersection) operation to generate column R_{max} is replaced by a weighted aggregation. The weight elements are the parameter which is associated with each goal element, respectively, to include the importance factor of individual goals to the business.

$$\text{The } R_{\text{aggregate}} = \text{OWA}(r_{i1} \ldots r_{im}) \text{ for } j = 1 \text{ to } m \tag{4}$$

Step 5 Perform intersection operation on column, $R_{\text{aggregate}}$ such that Decision $D = \text{Min } \{R_{max}\}$.

4.5   Illustrative Example for the Three Methods

In this section, we present the illustration of the three methods with the same data set used for all the three algorithms. The example data set is a representational data from the e-commerce domain. The attribute set is the set of the regions where the business is being considered for setup in the expansion plan. The goal set G is the set of following goals elements belonging to the e-commerce business domain [40]. The goal set G is the expected revenue generated in millions ($), based on the analyzed impact of each goal.

$G1$: Lack of competition (monopoly)
$G2$: Region-based strength of demand
$G3$: Regional buying capacity
$G4$: Revenue through advertisements
$G5$: Expensive substitutes
$G6$: Long-term vision of the firm
$G7$: Positive government business policy.

The data-element for the matrix could come from the market-survey and the expert system.

4.5.1 Max–Min Method

Step 1:

Attributes (Region)	G1	G2	G3	G4	G5	G6	G7
Region 1	20	22	18	15	25	33	18
Region 2	17	11	24	16	18	30	20
Region 3	29	21	33	25	18	28	22

Let us assume the largest value in the matrix to be 1 and least value to be 0.1 as the membership function mapping to fuzzify the data. The matrix with corresponding fuzzy values is as below

Attributes (Region)	G1	G2	G3	G4	G5	G6	G7
Region 1	0.60606061	0.66667	0.54545	0.45455	0.75758	1	0.54545
Region 2	0.51515152	0.33333	0.72727	0.48485	0.54545	0.90909	0.60606
Region 3	0.87878788	0.63636	1	0.75758	0.54545	0.84848	0.66667

$$\mu_D(A1) = \text{MIN} \{0.60606061, 0.66667, 0.54545, 0.45455,$$
$$0.75758, 1, 0.54545\} = \mathbf{0.45455}$$

$$\mu_D(A2) = \text{MIN}\{0.51515152, 0.33333, 0.72727, 0.48485,$$
$$0.54545, 0.90909, 0.60606\} = \mathbf{0.33333}$$

$$\mu_D(A3) = \text{MIN}\{0.87878788, 0.63636, 1, 0.75758, 0.54545,$$
$$0.84848, 0.66667\} = \mathbf{0.54545}$$

Hence

$$\mu_D(X_{\text{Optimal}}) = \text{MAX}\{0.45455, 0.33333, 0.54545\} = \mathbf{0.54545}$$

Hence The decision $D = \{\mathbf{Region\ 3}\}$.

4.5.2 Minimization of Regret Method (MMR)

Step 1:

Attributes (Region)	G1	G2	G3	G4	G5	G6	G7
Region 1	20	22	18	15	25	33	18
Region 2	17	11	24	16	18	30	20
Region 3	29	21	33	25	18	28	22

Step 2:

Attributes (Region)	G1	G2	G3	G4	G5	G6	G7
Region 1	20	22	18	15	25	33	18
Region 2	17	11	24	16	18	30	20
Region 3	29	21	33	25	18	28	22
Max	29	22	33	25	25	33	22

Step 3:

Attributes (Region)	G1	G2	G3	G4	G5	G6	G7
Region 1	9	0	15	10	0	0	4
Region 2	12	11	9	9	7	3	2
Region 3	0	1	0	0	7	5	0

Step 4:

Attributes (Region)	G1	G2	G3	G4	G5	G6	G7	R_{max}
Region 1	9	0	15	10	0	0	4	15
Region 2	12	11	9	9	7	3	2	12
Region 3	0	1	0	0	7	5	0	7

Step 5:

$$\mu_D(X_{Optimal}) = \text{MIN}\,(\mathbf{15}, \mathbf{12}, \mathbf{7}) = \mathbf{7}$$

So The decision $D = \{\mathbf{Region\ 3}\}$.

4.5.3 Generalization of MMR Method with OWA

Step 1, step 2 and step 3 remains the same as in MMR method
Step 4:

Attributes (Region)	G1	G2	G3	G4	G5	G6	G7	$R_{\text{aggregate}}$
Region 1	9	0	15	10	0	0	4	10.7
Region 2	12	11	9	9	7	3	2	12.55
Region 3	0	1	0	0	7	5	0	1.65
Weighting vector	0.4	0.2	0.3	0.1	0.1	0.15	0.4	

Step 5:

$$\mu_D(X_{\text{Optimal}}) = \text{MIN}\,(\mathbf{10.7, 12.55, 1.65}) = \mathbf{1.65}$$

So The decision D = {**Region 3**}.

4.6 Outliers in a Decision Table Data

We define outliers in a data set, for example in the attribute set or in the goal set as the element or very small proportion of elements which are substantially different from the large proportion of data. The substantial difference and large proportion could be defined quantitatively depending upon the nature of the data into consideration.

The presence of outliers might be treated as a special case, but their existence may result to conclude wrong decision. This deviation may be of varying degree for each algorithm.

The effect of outliers may depend on the nature of attributes and goals in the context of business and may also largely depend on the strategy of algorithm itself which we witness in Sect. 5.

The example e-commerce application that we used to illustrate the steps in Sect. 4.5 is reused again to analyze the impact of outliers with the changed data that has outlier elements to explain its impact on the decision D.

The attributes and the goals remain the same but the matrix elements are intentionally changed.

4.7 Illustrative Example for the Three Methods with Outlier Data Elements

In this section, we take the example of the data set for the attribute-goal matrix such that it contains outliers.

4.7.1 Max–Min Method

Attributes (Region)	G1	G2	G3	G4	G5	G6	G7
Region 1	6	7	7	9	8	18	15
Region 2	7	11	13	10	18	25	20
Region 3	5	21	33	25	7	28	22

If we observe the three rows of the matrix, we find that the data-element $G1$: Region 3 is 5 and $G5$:Region 3 are outlier to the highest degree. We still do not consider $G1$:Region 2 and $G4$:region 2 as outliers because they are not outlying with a very high degree (low value of membership).

Let us assume the largest value in the matrix to be 1 and least value to be 0.1 to fuzzify the data.

The matrix with corresponding fuzzy values would be as below

Attributes (Region)	G1	G2	G3	G4	G5	G6	G7
Region 1	0.18181818	0.21212	0.21212	0.27273	0.24242	0.54545	0.45455
Region 2	0.21212121	0.33333	0.39394	0.30303	0.54545	0.75758	0.60606
Region 3	0.15151515	0.63636	1	0.75758	0.21212	0.84848	0.66667

$$\mu_D(A1) = 0.18181818$$
$$\mu_D(A2) = 0.21212121$$
$$\mu_D(A3) = 0.15151515$$

Because, the algorithm functions on the intersection fuzzy operation in this step, the data element $G1$:region 3 induced inappropriate selection considering the definition of an optimal solution is to maximize gain with minimizing loss and hence attempt to minimize risk associated with this particular business. $G1$:region 3 which happens to be an outlier turns out to be a hot-spot data element which may discredit the rest of data for region 3 which forms the majority.

Hence, $\mu_D(X_{\text{Optimal}}) = 0.21212121$.

Hence The decision $D = \{$**Region 2**$\}$.

We analyze the same data matrix for the second algorithm in discussion in the next section and will try to find out if the decision made by the MMR algorithm is also Region 2.

4.7.2 Minimization of Regret Method (MMR)

Below are the steps on applying MMR algorithm on the data set containing outliers for which Max–Min algorithm decided for Region 2.

Step 1:

Attributes (Region)	G1	G2	G3	G4	G5	G6	G7
Region 1	6	7	7	9	8	18	15
Region 2	7	11	13	10	18	25	20
Region 3	5	21	33	25	7	28	22

Step 2:

Attributes (Region)	G1	G2	G3	G4	G5	G6	G7
Region 1	6	7	7	9	8	18	15
Region 2	7	11	13	10	18	25	20
Region 3	5	21	33	25	7	28	22
Max	7	21	33	25	18	28	22

Step 3:

Attributes (Region)	G1	G2	G3	G4	G5	G6	G7
Region 1	1	14	26	16	10	10	7
Region 2	0	10	20	15	0	3	2
Region 3	2	0	0	0	11	0	0

Step 4:

Attributes (Region)	G1	G2	G3	G4	G5	G6	G7	R_{max}
Region 1	1	14	26	16	10	10	7	26
Region 2	0	10	20	15	0	3	2	20
Region 3	2	0	0	0	11	0	0	11

Step 5:

$$\mu_D(X_{Optimal}) = \text{MIN}(26, 20, 11) = 11$$

So The decision $D = \{\textbf{Region 3}\}$.

The result is interesting as we found a deviation in the decision with this algorithm. The MMR method still picked the best possible solution because it considers regret matrix. And in step 4 functions with Min operator for the column represented by R_{max}. So $G1$:Region 3 no more is a hot-spot that could affect the decision because of the way algorithm works.

Now we try to analyze the situation with new data matrix with MMR while we have the outliers and that outlier happens to be hot-spot. These outliers affect the algorithm's step while it chooses inappropriate resulting considering the definition of the optimal solution remains intact that is, to maximize gain with minimizing loss and hence attempt to minimize risk associated with this particular business.

Now consider this data and apply the algorithm once again.

Step 1:

Attributes (Region)	G1	G2	G3	G4	G5	G6	G7
Region 1	6	7	7	9	8	18	15
Region 2	1	11	23	15	9	19	13
Region 3	5	21	33	25	7	28	22

Step 2:

Attributes (Region)	G1	G2	G3	G4	G5	G6	G7
Region 1	6	7	7	9	8	18	15
Region 2	1	11	23	15	9	19	13
Region 3	5	21	33	25	7	28	22
Max	7	21	33	25	18	28	22

Step 3:

Attributes (Region)	G1	G2	G3	G4	G5	G6	G7
Region 1	1	14	26	16	10	10	7
Region 2	6	10	10	10	9	9	9
Region 3	2	0	0	0	11	0	0

Step 4:

Attributes (Region)	G1	G2	G3	G4	G5	G6	G7	R_{max}
Region 1	1	14	26	16	10	10	7	26
Region 2	6	10	10	10	9	9	9	10
Region 3	2	0	0	0	11	0	0	11

Step 5:

$$\mu_D(X_{Optimal}) = MIN\,(\mathbf{26}, \mathbf{10}, \mathbf{11}) = 10$$

So The decision $D = \{$**Region 2**$\}$.

Now the point to consider is that the Region 3 has all of its elements having either 0 as regret value or a value 2, that is very close to 0 ($G1$:Region 3) except for $G5$:region 3 which appears to be a lately induced outlier and a hot-spot. Thus, this hot-spot results in an improper decision in the context of the optimal solution assumed of this business.

In next section, we try this data with Generalization of MMR method with OWA and try to analyze its impact on decision-making. One thing to note here is that both the algorithm that is MMR and works Generalization of MMR method with OWA works on regret matrix.

4.7.3 Generalization of MMR Method with OWA

The first three steps for the above data matrix remains the same.

We proceed with the fourth step where this algorithm differs by calculating $R_{aggregate}$ based on the weighting factor data. The weighting factor is not the attribute rather weights associated with each goal.

Step 4:

Attributes (Region)	$G1$	$G2$	$G3$	$G4$	$G5$	$G6$	$G7$	$R_{aggregate}$
Region 1	9	0	15	10	0	0	4	16.7
Region 2	12	11	9	9	7	3	2	14.25
Region 3	0	1	0	0	7	5	0	1.9
Weighting vector	0.4	0.2	0.3	0.1	0.1	0.15	0.4	

Step 5:

$$\mu_D(X_{Optimal}) = MIN(\mathbf{16.7}, \mathbf{14.25}, \mathbf{1.9}) = 1.9$$

So The decision $D = \{$**Region 3**$\}$.

This algorithm even in the presence of outlier is not impacted because the outlier is not a hot-spot that could affect the decision. It is because of the fact that this method is not only based on weighting vector but also on the principle of aggregation in step 4, instead of Max operation as in MMR. So the aggregation operation normalizes the anomaly.

Just for analysis, even if we keep the weighting factor equal that is 0.1 that is we neutralize the weighting factor we get the same decision which reasons that aggregation operation is the normalizing factor of the outlier's hot-spot effect. As we see below in step 5.

Step 4:

Attributes (Region)	G1	G2	G3	G4	G5	G6	G7	$R_{\text{aggregate}}$
Region 1	1	14	26	16	10	10	4	8.1
Region 2	6	10	10	10	9	9	9	6.3
Region 3	2	0	0	0	11	0	0	1.3
Weighting vector	0.1	0.1	0.1	0.1	0.1	0.1	0.1	

Step 5:

$$\mu_D(X_{\text{Optimal}}) = \text{MIN}\,(8.1, 6.3, 1.3) = 1.3$$

So The decision still remains the same that is $D = \{\textbf{Region 3}\}$.

5 Analysis of the Study and the Result

The above study of the effect of outliers suggests these points.

a. The effect of outliers in making the wrong decision is limited by the fact that they must be the hot-spot for any of the algorithm's critical steps.
b. Although the third method discussed,i.e., Generalization of MMR method with OWA appears to be already normalized of outliers, the choice of an algorithm does not just depend on the way it works. A business may prefer Max–Min algorithm as it is more aggressive towards risk accepting. There may be a reason for the business to take more risk provided they expect high gain which is a normal strategy in trading applications and hence our prior definition of optimal solution that is based on "to maximize gain with minimizing loss and hence attempt to minimize risk" becomes invalid.
In such situation the occurrence of outliers if they happen to be hot-spot worsen the situation while the strategist may still be willing to go with the Max–Min method. Hence we would need to normalize the effect of outliers as a preprocessing step of the impacting step of the algorithm. The outline of the preprocessing algorithm is suggested as below.
Algorithm:

Step 1: Check against predefined equation with factors including criticality threshold of outlier and the non-acceptable percentage of outliers and move to corrective steps starting from step 2 if the condition results false.
Step 1a: Check for occurrence of outliers.
Step 1b: Identify the outliers and find if it is a hot-spot for the method step in question.
Step 1c: Normalize the impact by modifying the suspected outliers by adding or subtracting an appropriate threshold constant.
Step 2: Proceed to the step of the decision-making method.

6 Learning and Self-evaluation Module

The learning and self-evaluation module of a system performs one of the two common baseline functions. One is to improve the decision-making ability and the other is to cross-check the decision being made at a particular instance against past recorded history for any projected deviation [41]. The two common baseline functionalities are analysis and reporting that run on "operational change history" or "log" database. The whole learning process should be aimed as evolving in nature on a regular basis with bare minimum manual intervention.

The analysis is about understanding the reason behind the past incidents and prediction is about speculating the future impact. Reporting, in particular, is rather about identifying the incidents affecting business and giving them a presentable form that could be easy to infer.

The learning and self-evaluation component is presented as an independent system in a separate region hence must be decoupled from the BI workflow or "application components region" to emphasize upon the need of its development as a pluggable framework to any such BI workflow system.

A more clear understanding of the flow of information is given below in Fig. 5. The flow below should be referred in the context of the big picture presented in Fig. 3 about the discussion of CRM application design with BI workflow and learning and self-evaluation systems.

In the figure above, the BI workflow feed input that is, the action and the activity log to the learning and self-evaluation system after collecting action from the fuzzy inference system. Whereas learning and self-evaluation system provides the rule modifications to fuzzy inference system after self-evaluation routine is invoked.

7 Conclusion and Future Prospects

We have focused on the problem of decision-making with vague data in the presence of outliers. We concluded that the outliers may result in incorrect decisions if the outlier has a strong effect on the important step. We call this outlier as hot-spot outlier. We attempted to distinguished algorithms for their capability to normalize the impact of hot-spot outliers with an emphasis that irrespective of this capability an algorithm may still be the desired one. So we proposed an outline of the algorithm that may be incorporated as preprocessing step to the affected step of decision-making algorithm to normalize the impact and enabling it to take a more appropriate decision.

For the study purpose, we have considered the three well-known algorithms. Exhaustive design of algorithm may be required for the preprocessing steps for each method in particular and all such possible methods of decision-making which are proposed in the literature in the context of evaluation for the effect of hot-spot outliers. There are numerous algorithms suggested by the various researchers which

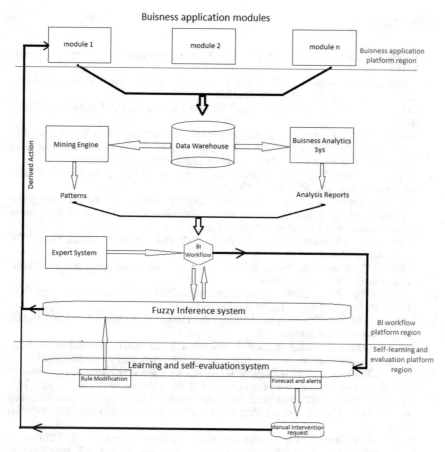

Fig. 5 Integrated design of learning and self-evaluation system

each individually, are subject to study for anomaly due to a presence of hot-spot outliers. One such algorithm which is yet another extension by Yager himself, which could be considered for such study is based on induced OWA [34] and another algorithm which is a rough set variation of MMR which is mainly proposed for clustering categorical data is also a case of study for the effect of outliers [42].

Apart from this core work we have also presented a detailed design-aspects of web-based application where decision-making and fuzzy logic has an important place and finally proposed a novel design for the cloud base solution as discussed in Sect. 4.1 with respect to Fig. 4. This proposed design could be of value to major cloud service providers as it suggests a separate and highly decoupled product that is "Fuzzy operation and Service Center". The major cloud service providers like Amazon or Microsoft have the infrastructure and the capability to build and offer such system as a cloud service offering just the way they do for analytics, big data and AI cloud service-based products.

References

1. Pawlak Z (1982) Rough sets. Int J Comput Inf 1–51. https://doi.org/10.1007/978-94-011-3534-4
2. Pawlak Z (1998) Rough set theory and its applications to data analysis. Cybern Syst 29:661–688. https://doi.org/10.1080/019697298125470
3. Zadeh LA (1965) Fuzzy sets. Inf Control 8:338–353. https://doi.org/10.1016/S0019-9958(65)90241-X
4. Zadeh LA (1978) Fuzzy sets as a basis for a theory of possibility. Fuzzy Sets Syst 1:3–28
5. Dubois D, Prade H (2007) International journal of systems science operations on fuzzy numbers. Int J Syst Sci 9:613–626. https://doi.org/10.1080/00207721.2010.492663
6. Bergeron P, Hiller CA (2005) Competitive intelligence. Annu Rev Inf Sci Technol 36:353–390. https://doi.org/10.1002/aris.1440360109
7. Ali A, Rakus-Andersson E (2009) Fuzzy decision making in business intelligence. Application of fuzzy models in retrieval of optimal decision. School of Engineering Blekinge Institute of Technology, Sweden, Thesis no:2009-6
8. Burns A (2008) Business intelligence. AMCIS 2008 Proc 11:161. https://doi.org/10.1007/978-3-642-36318-4
9. Poole DL, Mackworth A, Goebel RG (1998) Computational intelligence and knowledge. Comput Intell A Log Approach 1–22
10. Rahwan I, Simari GR (2009) Argumentation in artificial intelligence. Argum Artif Intell 171:1–493. https://doi.org/10.1007/978-0-387-98197-0
11. Johnson R (2005) J2EE development frameworks. Comput (Long Beach Calif) 38:107–110. https://doi.org/10.1109/MC.2005.22
12. Sun CT (1994) Rule-base structure identification in an adaptive-network-based fuzzy inference system. IEEE Trans Fuzzy Syst 2:64–73. https://doi.org/10.1109/91.273127
13. Bagis A, Karaboga D (2004) Artificial neural networks and fuzzy logic based control of spillway gates of dams. Hydrol Process 18:2485–2501. https://doi.org/10.1002/hyp.1477
14. Togai M, Watanabe H (1986) A VLSI implementation of a fuzzy-inference engine: toward an expert system on a chip. Inf Sci (Ny) 38:147–163. https://doi.org/10.1016/0020-0255(86)90017-4
15. Singh S, Karwayun R (2010) A comparative study of inference engines. In: ITNG2010—7th international conference on information technology new generations 53–57. https://doi.org/10.1109/ITNG.2010.198
16. Chen MS, Han J, Yu PS (1996) Data mining: an overview from a database perspective. IEEE Trans Knowl Data Eng 8:866–883. https://doi.org/10.1109/69.553155
17. Fan W, Bifet A (2013) Mining big data: current status, and forecast to the future. ACM SIGKDD Explor Newsl 14:1–5. https://doi.org/10.1145/2481244.2481246
18. Samad A (2017) Available online at www.ijarcs.info. Assessment of the prominent open source and free data mining tools, vol 8, pp 2058–2062
19. Chen H, Chung W, Xu JJ et al (2004) Crime data mining : a general framework. IEEE Comput Soc 50–56
20. Monaro RM, Vieira JCM, Coury DV, Malik OP (2015) A novel method based on fuzzy logic and data mining for synchronous generator digital protection. IEEE Trans Power Deliv 30:1487–1495. https://doi.org/10.1109/TPWRD.2014.2372007
21. Kohavi R, Rothlender N, Simoudis E (2002) Emerging Trends in Business Analytics. Commun ACM 45:45–48. https://doi.org/10.1016/S0026-0657(02)80552-5
22. Zhenning Y, Vijayashree J, Jayashree J (2017) Fuzzy logic based diagnosis for liver disease using CBC (Complete Blood Counts). J Comput Math Sci 8:202–209
23. Liuyang Z, Jayashree J, Vijayashree J (2017) Fuzzy logic based thyroid diagnosis system. J Comput Math Sci 8:218–224
24. Kai W, Vijayashree J, Jayashree J (2017) Fuzzy logic based diagnosis system for mental illness. J Comput Math Sci 8:210–217

25. Amiri FM, Khadivar A (2017) A fuzzy expert system for diagnosis and treatment of musculoskeletal disorders in wrist. Teh Vjesn Tech Gaz. https://doi.org/10.17559/TV-20150531034750
26. Liu JF (2007) Yu DR (2007) A weighted rough set method to address the class imbalance problem. Proc Sixth Int Conf Mach Learn Cybern ICMLC 7:3693–3698. https://doi.org/10.1109/ICMLC.2007.4370789
27. Chen IJ, Popovich K (2003) Understanding customer relationship management (CRM) people, process and technology. Bus Process Manage J 9:672
28. An M, Chen Y, Baker CJ (2011) A fuzzy reasoning and fuzzy-analytical hierarchy process based approach to the process of railway risk information: a railway risk management system. Inf Sci (Ny) 181:3946–3966. https://doi.org/10.1016/j.ins.2011.04.051
29. Cho YH, Kim JK (2004) Application of web usage mining and product taxonomy to collaborative recommendations in e-commerce. Expert Syst Appl 26:233–246. https://doi.org/10.1016/S0957-4174(03)00138-6
30. Ho GTS, Ip WH, Wu CH, Tse YK (2012) Using a fuzzy association rule mining approach to identify the financial data association. Expert Syst Appl 39:9054–9063. https://doi.org/10.1016/j.eswa.2012.02.047
31. Benatallah B, Motahari Nezhad H (2008) Service oriented architecture: overview and directions. Adv Softw Eng 5316:116–130. https://doi.org/10.1007/978-3-540-89762-0_4
32. Dhull VK (2017) An optimized deployment approach for intercloud and hybrid cloud. Int J Adv Trends Comput Sci Eng 9:266–274
33. Weber A, Thomas R (2005) Key performance indicators—measuring and managing the maintenance. IAVARA Work Smart 1–16. https://doi.org/10.1016/j.ecolind.2009.04.008
34. Yager RR, Filev DP (1999) Induced ordered weighted averaging operators. IEEE Trans Syst Man Cybern Part B Cybern 29:141–150. https://doi.org/10.1109/3477.752789
35. Allende-Cid H, Canessa E, Quezada A, Allende H (2011) An improved fuzzy rule-based automated trading agent. Stud Inf Control 20:135–142. https://doi.org/10.1109/SCCC.2010.33
36. Yager RR (2004) Decision making using minimization of regret. Int J Approx Reasoning 36:109–128. https://doi.org/10.1016/j.ijar.2003.10.003
37. Faruque Uddin M (2012) Application of fuzzy logic in sociological research: an instance of potential payoff. Bangladesh eJ Soc 9:227–237. https://doi.org/10.19044/esj.2017.v13n5p227
38. Plasmas CL (2017) Faruque Uddin. IEEE Trans Fuzzy Syst 35:1194–1195. https://doi.org/10.1109/TFUZZ.2017.2682542
39. Siddique M (2009) Fuzzy decision making using max-min method and minimization of regret method (MMR), pp 1–23
40. Search C (2017) Factors that affect the profitability of firms, pp 1–5. http://www.economicshelp.org/microessays/profit/
41. Kahkeshan B, Hassan SI (2017) Assessment of accuracy enhancement of back propagation algorithm by training the model using deep learning. Orient J Comput Sci Technol 10:298
42. Parmar D, Wu T, Blackhurst J (2007) MMR : an algorithm for clustering categorical data using rough set theory. Data Knowl Eng 63:879–893. https://doi.org/10.1016/j.datak.2007.05.005

GSA-CHSR: Gravitational Search Algorithm for Cluster Head Selection and Routing in Wireless Sensor Networks

Praveen Lalwani, Haider Banka and Chiranjeev Kumar

1 Introduction

Wireless sensor networks (WSNs) is the collection of several thousands of sensor nodes. These nodes operate in a cooperative manner in order to perform the task, basically for data communication. Each sensor node collects data from its surroundings within its predefined sensing range. Afterwards, it processes the data and transmits to the positioned base station (BS) [1–3]. Every sensor node has equipped with non-replaceable energy supply and dissipates energy while transmitting, receiving and processing. Therefore, to reduce energy consumption in WSNs is one of the most critical issues [4]. There are several applications of WSNs, such as temperature monitoring, home automation, traffic monitoring, disaster warning systems, health care, etc.

WSN is partitioned into several small groups called clusters. Each cluster has a leader node known as CH. Clustering provides inherent minimization of energy consumption [5–7], it supports scalability as well as minimizes the work of the BS by making local decisions. Improper CH selection may lead to quick death of the CHs and degrades the overall performance of the network [8–10]. In addition, the computational complexity of finding optimum sensor nodes as CH nodes for a large-scale WSN is exponential in nature, most of the brute force algorithms cannot capture the explosive growth with size [11].

P. Lalwani · H. Banka (✉) · C. Kumar
Indian School of Mines, Dhanbad, India
e-mail: haider.banka@gmail.com

P. Lalwani
e-mail: praveenlalwani@cse.ism.ac.in

C. Kumar
e-mail: kumar.c.cse@ismdhanbad.ac.in

© Springer Nature Singapore Pte Ltd. 2017
R. Ali and M. M. S. Beg (eds.), *Applications of Soft Computing for the Web*,
https://doi.org/10.1007/978-981-10-7098-3_13

After cluster formation, every CH transmits the captured data to the BS, but the direct transmission is impractical in large WSN. The reason behind that CH transmits data to the BS when it lies in within the communication range, but this scenario is not possible in large WSN. Therefore, relay nodes are selected between CH and the BS for enabling the communication (i.e., routing) [12, 13]. In hierarchical routing, CH receives data from several sources, such as intercluster and intra-cluster members, it performs aggregation and transfers a single combined information packet from lower cluster layer to the higher until it reaches to the BS as shown in Fig. 1. It is worthwhile to note that the computational complexity of finding an optimum route for a large-scale WSN is very high because complexity increases exponentially with the network size and cannot be captured by brute force approaches [14]. Therefore, many researchers are focusing nature inspired techniques (GA, PSO, and many others) to provide the approximate solution of above stated problems, i.e., CH selection and routing.

Gravitational search algorithm (GSA) is one of the nature inspired approaches devised by Rashedi et al. [15]. It is based on the law of gravity and motion. In GSA, all objects attract each other by force of gravity and the resultant movement of objects toward objects with heavier masses. Since, GSA has provided better performance than well-established techniques namely, RGA and PSO in most of the cases [15], it motivated us to a proposed GSA-based algorithm for both CH selection and routing in WSNs. To the best of our knowledge, it is the first such attempt.

In this study, we have proposed two algorithms based on the GSA and one novel potential function. The first algorithm is CH selection and other is routing. Our first algorithm efficiently selects the near optimal sensor nodes as CHs out of all in such a way that energy consumption is minimized in the network. In order to achieve the above stated objective, the new fitness function is proposed and an effective encoding scheme is also formulated. Afterwards, non-CH nodes assigned to the CH nodes using our derived potential function. Finally, the routing algorithm is devised to find the near optimal route from each CH to the BS. To accomplish this,

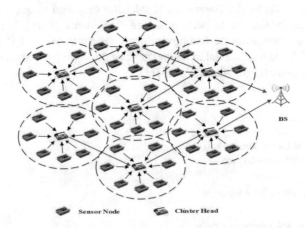

Fig. 1 Cluster head communication to the base station

a mapping scheme, a new fitness function, and an efficient encoding scheme are devised. In the extensive simulation, firstly our proposed work (GSA-CHSR) is compared with some recent existing approaches. Afterwards, we have also executed well-established algorithms GA and PSO on the above stated problems and compared with the proposed work in terms of convergence rate and performance metrics as well.

The summary of the proposed work are as follows:

- GSA-based cluster head selection algorithm based on the multiple objectives, such as the residual energy of CH, intra-cluster distance, and CH balancing factor.
- An efficient potential function for assigning sensor nodes to the CH.
- GSA-based routing algorithm with a novel fitness function which also composed of multiple objectives namely, residual energy and distance. It provides the complete routing solution for multi-hop communication between each CH and the BS.
- In performance analysis, the proposed algorithm is compared with some of the existing techniques to show the superiority over them.

The rest of the work is organized as follows. Next section shows literature review related to our proposed algorithm in brief. The system model and terminology are discussed in Sect. 3. The proposed CH selection algorithm, potential function derivation, and routing algorithm are discussed in detail in Sect. 4. Experimental analysis with some existing algorithms is shown in Sect. 5. Finally, the paper is concluded in Sect. 6.

2 Literature Review

In the section, most recent works in clustering and routing algorithms based on the brute force approaches and nature inspired techniques is presented.

2.1 Clustering

A wide variety of clustering algorithms are available in the literature [16–19]. Among these, LEACH is a very popular one [19]. It takes a local decision to select CHs among all sensor nodes. It includes randomized rotation of CHs to evenly distribute the energy consumption over the network. However, CH with low residual energy have the possibility of selection, it increases the energy consumption of low energy nodes. Heinzelman et al. [16] proposed LEACH-Centralized (LEACH-C) protocol. Initially, all the nodes send its location and residual energy to the BS, and then, it computes the average residual energy of the network. Thereafter, it forms the cluster using a simulated annealing algorithm. However, node degree in cluster formation has not been taken into consideration. Wang et al. [18] improves the performance of LEACH by taking residual energy into consideration in selection probability of CH. But, in large WSN direct communication between CH and the BS not possible. Chang and Ju [20] proposed a energy saving clustering architecture. It takes two parameters as input in cluster formation process, i.e., average distance from sensor nodes to the BS, and center point of the WSN. Afterwards, CHs are selected based on the residual energy. Finally, in this way load on every cluster is properly balanced. Therefore, network life increases. Yang and Ju [21] proposed a centralized algorithm. Its aim is to provide adaptive intra-cluster multi-hop approach to reduce the transmission power of all the sensor nodes. In this mechanism, CHs are selected based on the location and residual energy information. Moreover, tree structure is formed within the cluster. However, it does not take care of residual energy for tree structure formation. Bagci and Yazici [22] proposed a distributed algorithm. In competitive radii calculation, fuzzy logic was used which takes energy and distance to the BS as input. Therefore, it reduces the energy consumption in the network. Lee and Cheng [23] proposed a distributed algorithm. Initially, it uses LEACH algorithm for CH selection. Afterwards, fuzzy logic is used over selected CHs to calculate the candidate chance which considers residual energy and expected residual energy as input. If the chance of candidate is bigger than the remaining, then it becomes CH. However, it is does not take care of distance to the BS and node density. Kumar et al. [24] proposed a fuzzy logic-based clustering algorithm. It selects the CH's by fuzzy inference system using distance, node density, and battery level. The obtained output is the CH selection probability. Therefore, it increases the life of the network. However, it is unable to provide adaptive multi-hop communication.

2.2 Routing

In large WSNs, direct communication between sensors nodes or CHs and the BS may not possible due to limited transmission range of nodes. Every node needs to communicate with the BS at the minimum cost. Thus, the routing protocols have been proposed by investigators [17, 25, 26]. Among these, HEED is a very popular one [17]. It selects CH based on the residual energy, whenever, the residual energy of two or more CHs is similar, then the AMRP cost function is used to select one of them. Every CH communicates directly to the BS or in multi-hop fashion based on the transmission distance between CH and the BS. Senouci et al. [27] proposed the modified version of HEED. Its cluster formation process is similar to HEED. But, it needs extra time to establish the clusters due to lots of messages are exchanged between non-CH nodes. It extends the functionality of HEED by dividing cluster into small zones based on the transmission range between non-CH and CH nodes. Therefore, it enhances the life of the network. Lai et al. [28] proposed an unequal cluster formation mechanism. Clusters nearer to the BS are smaller in size and larger when it far from the BS. Therefore, load on every CH is balanced that protects from cluster reconfiguration. Moreover, it also increases the network lifetime as well as stability. Abdulla et al. [29] proposed hybrid routing protocol. It performs flat routing inside the hot-spot zone and hierarchical outwards. However, it does not consider the effect of the hybrid boundary on network lifespan. Yu et al. [30] proposed a routing protocol (EADC). It selects the next-hop using energy and node degree. However, it does not take care of transmission distance for next-hop selection. Maryam and Reza [31] enhance the performance of EADC by adding one more parameter for its next-hop selection process, i.e., transmission power. Another routing algorithm was proposed by Song and Cheng-lin [32]. Initially, the competitive radii for every cluster and chance for its respective CH is calculated using fuzzy logic, which considers density, distance, and energy of the node as input. Thereafter, the routing path is established for every CH using ant colony optimization (ACO). Therefore, it enhances the network performance. Bhari et al. [33] proposed a genetic-based routing algorithm. Its objective is to maximize the lifetime of the network. However, its objective function does not take care about residual energy, node degree, and distance to the BS. Elhabyan and Yagoub [34] proposed PSO-based clustering and routing protocols. In clustering process, a potential function consists of energy, cluster quality, and network coverage maximization. In routing process, an objective function consists of energy and link quality maximization. However, it does not take care of power control for next-hop selection.

2.3 Advantages of Proposed Method Over Existing Methods

- Brute force approaches mentioned in the literature (refer to Sects. 2.1 and 2.2) are not able to solve CH selection and routing problem for large-scale WSNs [18, 29, 30]. Because both problems have been proven to be NP-hard in nature by investigators [11, 14].
- Existing meta-heuristic approaches mentioned are not able to provide promising solution due to lack of consideration of essential factors in the formulation of the fitness function [34], whereas, in the proposed approach, GSA is used with proper consideration of essential parameters namely, energy, distance, and CH balancing factor. So, that promising solution was achieved compared to existing approaches.
- We have derived efficient encoding schemes for the CH selection and routing algorithms in contrast to existing encoding schemes [33].
- In the cluster formation process, non-CH sensor nodes join a CH based on the derived novel potential function. In contrast, the existing algorithms [33, 34] where non-CH sensor nodes join the CH based on the distance only, which may cause imbalance load of the CHs and may lead to faster energy depletion of the network.

3 Preliminaries

The current section consists of abbreviations, energy model, network model, and description of GSA.

Table 1 Description of notations/abbreviations

Notation	Description
E_{T_x}	Energy consumption during transmission
E_{R_x}	Energy consumption during receiving
E_{amp}	Energy consumption during amplification
E_{da}	Aggregation energy
E_{fs}	Energy consumption when free space model is used
E_{mp}	Energy consumption when multi-path fading channel model is used
$d_{(non\text{-}CH,CH)}$	Distance between non-CH and CH
$d_{CH\text{-}BS}$	Distance between CH and BS
$M_{a,j}$	Active gravitational mass of jth particle
$M_{p,i}$	Passive gravitational mass of ith particle
f_{ij}	Force between agent i and j
E_{CH_j}	Residual energy of current cluster head CH_j, where $j \varepsilon [1, m]$
$dis(s_i, s_j)$	Distance between two sensor nodes i and j
S	Denotes the set of sensor nodes
C	Denotes the set of cluster head nodes
I_j	Number of intra-cluster members
NH	The next-hop of current CH
m	Denotes number of CHs
n	Denotes number of sensor nodes
ζ_i	Denotes routing path for ith CH
N_p	Number of agents
A_i	Represents ith agent
$m_i(t)$	Mass of ith agent at time t
m_{ii}	Inertial mass of ith agent
$R_{i,j}$	Euclidean distance between agent i and j
T	Maximum number of iterations
E_{elec}	Energy consumed in transmitter circuitry
r	Random number between 0 and 1
RN_i	Real number between 0 and 1
▷	Represents the comment

3.1 Notation

The notations and abbreviations used in this study are shown in Table 1.

3.2 Energy Model

In data processing, sensor node depletes its energy which is calculates using first-order radio model [19]. Energy consumption for transmitting L bits of data at communication distance d_t is shown in Eq. 3.1, where E_{elec} is the energy consumed in transmitter circuitry and E_{amp} is the energy required for amplification.

$$E_{Tx} = \begin{cases} E_{elec} * L + E_{amp} * L * d_t^2 \text{ if } d_t < d_o; \\ E_{elec} * L + E_{amp} * L * d_t^4 \text{ if } d_t \geq d_o. \end{cases} \tag{3.1}$$

Energy consumption for receiving L bits of data is shown in Eq. 3.2

$$E_{R_x} = E_{elec} * L \tag{3.2}$$

3.3 Network Model

In this paper, two-tired wireless sensor network is considered with following properties. The sensors are stationary after random deployment and distance calculation is based on the received signal strength. In node configuration, all nodes are homogeneous and have equal capabilities for processing and communication. The number of CHs are m, number of sensors are n and one BS which is placed at various locations, i.e., center, corner, and outside the network. We assume that every sensor belongs to only CH, they perform sensing periodically and send data to their respective CH. Initially, sensor nodes broadcast residual energy and location information toward the BS. Afterwards, CH selection and routing schedule is computed by our proposed algorithms at BS without any power constraint.

3.4 Gravitational Search Algorithm

Gravitational search algorithm (GSA) is one of the nature inspired techniques [15]. It is used to find the approximation solution of NP-hard problems. It was devised by E. Rashedi et al. in 2009, based on the law of gravity.

GSA is based on the principle of the Newtonian law of gravity and motion: "*Every particle in the universe attracts every other particle with a force that is*

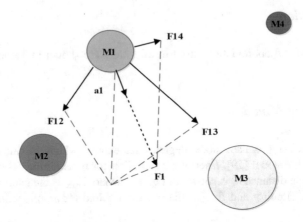

Fig. 2 Every mass (agent) accelerate toward the result force

directly proportional to the product of their masses and inversely proportional to the square of the distance between them". In GSA, agents are assumed as objects and their performance is evaluated by their masses. All these objects/agents attract each other by the gravity force as shown in Fig. 2. Each agent has four specifications, i.e., position, active gravitational mass, passive gravitational mass, and inertial mass.

3.4.1 Position

In GSA, each agent represents the potential solution of the considered problem. In the initialization process, each dimension of an agent is randomly initialized (Initialization for CH selection and routing problems in WSN refer to Sects. 4.1.2 and 4.3.2). Let ith agent is shown in Eq. 3.3.

$$A_i = (A_i^1, A_i^2, A_i^3, \ldots, A_i^d, \ldots, A_i^n) \quad \text{for } i = 1, 2, \ldots, N. \tag{3.3}$$

where, A_i^d represents the position of ith agent in dth dimension.

3.4.2 Active Gravitation Mass

It measures the strength of the gravitational field applied by a particular object. The gravitational field of an object with small active gravitational mass is weaker compared to larger one. Gravitational mass is analogous to the fitness value.

3.4.3 Passive Gravitation Mass

It measures the strength of object interaction with the gravitation field. In the gravitational field, an object with a smaller passive gravitational mass experiences a smaller force than a larger. It is directly proportional to the fitness value.

3.4.4 Mapping Scheme

Mapping scheme to estimate mass from the fitness value is shown in Eq. 3.4.

$$m_i(t) = \frac{\text{fit}_i(t) - \text{worst}(t)}{\text{best}(t) - \text{worst}(t)} \tag{3.4}$$

where $\text{fit}_i(t)$ is ith agent fitness, $\text{worst}(t)$ is minimum fitness value, and $\text{best}(t)$ is maximum fitness value at time t. As we know mass $m_i(t)$ cannot be zero, it is adjusted accordingly by adding small value ϵ in Eq. 3.4 as shown in Eq. 3.5.

$$m_i(t) = \frac{\text{fit}_i(t) - \text{worst}(t) + \epsilon}{\text{best}(t) - \text{worst}(t)} \tag{3.5}$$

3.4.5 Force

The force acting on mass i from mass j in dth dimension at time t is shown in Eq. 3.6, where $R_{i,j}(t)$ is euclidean distance between two agents i and j, $M_{p,i}$ is passive gravitational mass and $M_{a,j}$ is active gravitational mass.

$$f_{ij}^d(t) = \frac{G(t) * M_{p,i}(t) * M_{a,j}(t)}{R_{i,j}(t)} (A_j^d(t) - A_i^d(t)) \tag{3.6}$$

3.4.6 Acceleration

Its calculation is shown in Eq. 3.7, where M_{ii} is the inertial mass of ith agent which measures the resistance on an object when gravitational force is applied on it.

$$a_i^d(t) = \frac{f_i^d(t)}{M_{ii}} \tag{3.7}$$

3.4.7 Velocity and Position Update

After generation of an acceleration value, each dimension of an agent is updated using velocity and position Eqs. (3.8 and 3.9), respectively.

$$V_i^d(t+1) = r * V_i^d(t) + a_i^d(t) \tag{3.8}$$

$$A_i^d(t+1) = A_i^d(t) + V_i^d(t+1) \tag{3.9}$$

where, r is a random number in the interval [0, 1], V_i is velocity, A_i^d is position and a_i^d is an acceleration of ith agent in dth dimension at time t.

3.4.8 K_{best} Calculation

The performance of GSA can be enhanced by controlling of exploration and exploitation. In order to achieve the aforementioned purpose, K_{best} the function is used as shown in Eqs. (3.10 and 3.11).

$$K_{best} = Final\_per + (1 - iteration/T) * (100 - Final\_per); \tag{3.10}$$

$$K_{best} = round(N_p * K_{best}/100) \tag{3.11}$$

where, T is maximum iterations and N_p is the number of agents.

4 The Proposed Model

The proposed model composed of two derived algorithms and one potential function. The first algorithm takes care of CH selection, whereas, another addresses the routing process for each CH. The first algorithm efficiently selects an approximate set of sensor nodes as CHs among all sensor nodes (refer to Sect. 4.1). Thereafter, a novel potential function has been proposed to assign sensor nodes to the CH nodes (refer to Sect. 4.2). Finally, the routing algorithm is proposed to search the near-optimal path from every CH to the BS (refer to Sect. 4.3).

4.1 GSA-Based CH Selection Algorithm

This algorithm is devised to select the approximate set of sensor nodes as CHs based on the energy, distance, and CH balancing factor.

4.1.1 Representation of an Agent

In GSA, a potential solution is represented by an agent. In CH selection process, an agent represents a set of sensor nodes selected as CHs among all sensor nodes and dimension of each agent is equal to the number of CHs in the network.

4.1.2 Initialization of an Agent

Each dimension of an agent is initialized by random node_id and is mapped with its 2D coordinates. Let $A_i = (A_{i,1}(t), A_{i,2}(t), \ldots, A_{i,m}(t))$ be the ith agent, where, $A_{i,d}(t) = (x_a, y_b), \forall_a 1 \le a \le n, \forall_d 1 \le d \le m$ denotes the coordinates of CH.

Example 1 Let the size of network area is 200×200, number of nodes are 200 and number of CHs are 5% of total number of sensor nodes, i.e., 10. Therefore, each dimension of an agent is initialized with a random number between 1 and 200, i.e., node_id. Afterwords, it is mapped with its 2D coordinates as shown in Fig. 3.

4.1.3 Derivation of Fitness Function

The formulation of fitness function by considering energy, distance, and CH balancing factor are as follows:

(a) **Energy**: All CH members transmit data to their respective CH, then CH aggregates the received data and it converts into the single transmission packet. Afterwards, packet forwards toward the BS. In this way, CH consumes more energy compared to the remaining sensor nodes. Therefore, a sensor node with higher residual energy is a more prominent choice as a CH, it means that energy consumption is less for low energy nodes and more for high energy nodes. So, our first objective is f_1 which can be minimized as follows:

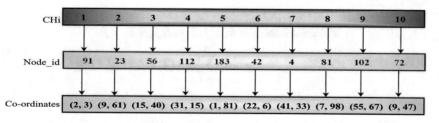

Fig. 3 Representation of an agent for CH selection

Objective 1

$$f_1 = \sum_{i=1}^{m} \frac{1}{E_{CHi}} \tag{4.1}$$

(b) **Distance**: It is the average distance between CH members and their respective CH. If it is minimum, it will decrease the energy consumption of low energy sensor nodes, i.e., CH members. This increases the number of alive nodes in the network. So, the second objective is f_2 which can be minimized as follows:

Objective 2

$$f_2 = \sum_{j=1}^{m} \left(\sum_{i=1}^{I_j} dis(s_i, CH_j)/I_j \right) \tag{4.2}$$

(c) **CH balancing factor**: Sensor nodes are assigned to the CH based on the minimum distance because it reduces the energy consumption of each sensor node during the transmission process. Due to the random deployment of sensor nodes, some CHs have large CH members and some have less. Therefore, CH which has large CH members demises rapidly. So, our third objective is to derive the balancing factor to balance the load on each CH. It will reduce the energy consumption of each CH and increases the lifespan of the network. Therefore, the third objective is f_3 which can be minimized as follows:

Objective 3

$$f_3 = \sum_{j=1}^{m} \left[\frac{n}{m} - I_j \right] \tag{4.3}$$

All of the above-mentioned objectives are not conflicting to each other. Therefore, weighted sum approach is applied to frame single objective function using multiple objectives, where, α_1, α_2 and α_3 are the weighted value assigned to each objective.

$$\text{Minimize Fitness} = \alpha_1 \times f_1 + \alpha_2 \times f_2 + \alpha_3 \times f_3,$$
$$\text{where, } \sum_{i=1}^{3} \alpha_i = 1; \text{ and } \alpha \in (0, 1) \tag{4.4}$$

4.1.4 Velocity and Position Update

In this step, every position of an agent is updated using velocity and position update Eqs. (8 and 9), respectively. The predefined size of the target area is 200×200. So, newly generated 2D coordinates should be in the range between 1 and 200. Due to the algebraic calculation, it may cause newly generated 2D coordinates may lie out of the target area. In that case, the random number generated between 1 and 200 (represents node-id) and its 2D coordinates are considered to replace the corresponding position (dimension).

4.1.5 Pseudo Code for Cluster Head Selection

Algorithm 1: GSA-Based Optimal CH Selection

Inputs: (1) Set of sensor nodes $S = (s_1, s_2, s_3, \ldots, s_n)$.
   (2) Predefined number of agents N_p.
   (3) Dimension of agent m.
   (4) Kbest=N_p.
Result: Set of near optimal cluster heads $C = (CH_1, CH_2, \ldots, CH_m)$.

 Step 1: Initialize the agent A_i, $\forall_i 1 \leq i \leq N_p$.
Step 2:
while *(i! = N_p)* **do**
 | Calculate $Fitness(A_i)$ ▷ using Eq. 4.4
end
Step 3:
while *(i! = N_p)* **do**
 | $best = maximum(Fitness(A_i))$
 | $worst = minimum(Fitness(A_i))$
end
Step 4:
while *(i! = N_p)* **do**
 | Calculate $mass(A_i)$ ▷ using Eq. 3.4
end
Step 5:
while *(s! = T)* **do**
  while *(i! = N_p)* **do**
    while *(j! = Kbest)* **do**
     5.1 Calculate the force of agent A_i ▷ using Eq. 3.6
     5.2 Calculate the acceleration of agent A_i ▷ using Eq. 3.7
     5.3 update the cordinate of CH_i ▷ using Eqs. (3.8 & 3.9)
     5.4 update the Kbest value   ▷ using Eqs. (3.10) and (3.11)
    end
  end
end
Step 6: Assign sensor nodes to a cluster head using
 $minimum(fitness(A_i))$.

4.1.6 Illustrative Example of CH Selection Process Using GSA

Let the first position (dimension) of an active agent is (92, 12.5) with mass = 0.3. The first dimension of a passive agent is (76, 23.1) with mass = 0.8, $r = 0.5$ (random number), $T = 100$ (maximum number of iterations), euclidean distance between two agents (R) is 57, initially all the velocities are 0, and the number of sensor nodes are 200.

The first position of d_t the agent is updated as shown in Table 2. Afterwards, nearest sensor node coordinates to the updated position are considered to replace the corresponding position. In this way, each position of each agent is updated. Due to the algebraic calculation, it may cause newly generated position may not satisfy the range, i.e., negative or greater than 200. In that case, the random number is generated between 1 and 200 (node_id) and its coordinates are chosen to replace the position. Repeat all the steps iteratively, until the maximum number of iterations is reached.

Table 2 Procedure for updating ith agent in CH selection

Steps to update every dimension of ith agent are as follows:
Step 1: Estimate the gravitational constant $G(t)$, it will reduce with iterations to control the search accuracy
$G(t) = G_0 * e^{-\alpha*t/T}$, where $G_0 = 100$, $\alpha = 20$, $t = 1$
$G(t) = 100 * e^{-20*1/100} = 83.33$
Step 2: Calculate the Force_x and Force_y between the active and passive agents using Eq. 6. In the similar way, calculate the force of each dimension of each agent
Force_x $= \frac{G*M_a*M_b}{R+\varepsilon} * (x2 - x1) = ((83.33 * 0.3 * 0.8)/(57+0.1)) * (76 - 92) = -5.6138$
Force_y $= \frac{G*M_a*M_b}{R+\varepsilon} * (y2 - y1) = ((83.33 * 0.3 * 0.8)/(57+0.1)) * (23.1 - 12.5) = 3.508$
Step 3: Calculate the acceleration for first position of active agent a_1^1
Force_x$_1^{(1)} = \sum\limits_{j=1, j \neq 1}^{N_p}$ rand $*$ Force_x$_{ij}^1 = 17.2$ ▷ force towards x-axis
Force_y$_1^{(1)} = \sum\limits_{j=1, j \neq 1}^{N_p}$ rand $*$ Force_y$_{ij}^1 = 19.2$ ▷ force towards y-axis
$ax_1^1 = \frac{\text{Force\_x}_1^{(1)}}{M_{ii}} = \frac{-17.2}{0.3} = -57.33$ ▷ acceleration towards x-axis
$ay_1^1 = \frac{\text{Force\_y}_1^{(1)}}{M_{ii}} = \frac{19.2}{0.3} = 64$ ▷ acceleration towards y-axis.
Step 4: Update the velocity and position of first dimension of an agent along x_coordinate using Eqs. (8 and 9), respectively
$Vx_1^1 = r * V_1^1 + a_1^1 = 0.5 * 0 - 57.33 = -57.33$
$A_1^1(x) = A_1^1 + V_1^1 = 92 - 57.33 = 34.67$
Step 5: Updates velocity and position along y_coordinate
$V_1^1 = r * V_1^1 + a_1^1 = 0.5 * 0 + 64 = 64$
$A_1^1(y) = A_1^1 + V_1^1 = 12.5 + 64 = 76.5$
Step 6: Finally, the first position of an agent (92, 12.5) is updated to new position (34.67, 76.5)

4.2 Formulation of Potential Function for Cluster Formation

The potential function is derived for assigning sensor nodes to the CH. The parameters considered for the derivation are mentioned below:

(a) *Residual energy of CH*: The sensor node preferred to join to the CH that has higher residual energy, when compared against other CHs within its sensing range.

$$\text{Potential\_fun}(s_i) \propto \text{Energy}(CH_j) \tag{4.5}$$

(b) *Sensor node and CH distance*: The sensor node preferred to join to the CH, which has less transmission distance. So, that it saves the adequate amount of energy during the transmission process.

$$\text{Potential\_fun}(s_i) \propto \frac{1}{\text{Distance}(s_i, CH_j)} \tag{4.6}$$

All the aforementioned Eqs. (4.5 and 4.6) are combined to obtain the final potential function.

$$\text{Potential\_fun}(s_i) \propto \frac{\text{Energy}(CH_j)}{\text{Distance}(s_i, CH_j)} \tag{4.7}$$

$$\text{Potential\_fun}(s_i) = L \times \frac{\text{Energy}(CH_j)}{\text{Distance}(s_i, CH_j)} \tag{4.8}$$

where L is the proportionality constant. The sensor node (s_i) assigned to the CH_j which has high-potential function values.

4.3 GSA-Based Routing Algorithm

This algorithm is formulated to select near the optimal path from every CH to the BS using energy and distance.

4.3.1 Representation of an Agent

In routing process, each agent represents the data forwarding path from each CH to the BS and the dimension is equal to the number of CHs in the network.

4.3.2 Initialization of an Agent

Initially, each position of an agents is randomly initialized by a random number between 0 and 1 represented as RN_i and shown in Fig. 4.

Let $A_i = (A_{i,1}(t), A_{i,2}(t), A_{i,3}(t), \ldots, A_{i,m}(t))$ be the ith agent, where every position $A_{i,d} = (0, 1)$, $\forall_i 1 \leq i \leq N_p$, $\forall_d 1 \leq d \leq m$. Afterwards, the value of every $A_{i,d}$ is mapped with the next-hop (say CH_k). This shows that CH_d transmits data to CH_k. The mapping scheme is shown in Eq. 4.9.

$$f(x) = \left\{ i, \quad \text{for which} \left| \left(\tfrac{i}{k} - A_{i,j} \right) \right| \text{is minimum}, \quad \forall_i, 1 \leq i \leq k \right. \qquad (4.9)$$

where, k represents the number of next-hops within its sensing range and $A_{i,j}$ is the position of agent.

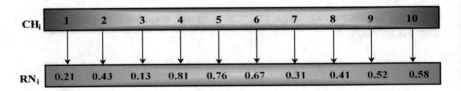

Fig. 4 Representation of ith agent in routing

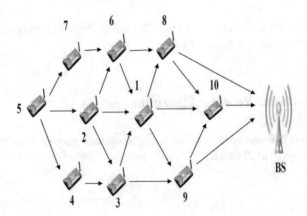

Fig. 5 Pictorial view of wireless sensor subnetwork

Table 3 CHs with their selected next-hop

Cluster Head (CH_i)	Feasible next-hop (CH_j)	Next-hop
Height 1.3pt CH_1	$\{CH_8, CH_9, CH_{10}\}$	CH_8
CH_2	$\{CH_1, CH_3, CH_6\}$	CH_3
CH_3	$\{CH_1, CH_9\}$	CH_1
CH_4	$\{CH_3\}$	CH_3
CH_5	$\{CH_2, CH_4, CH_7\}$	CH_7
CH_6	$\{CH_1, CH_8\}$	CH_8
CH_7	$\{CH_6\}$	CH_6
CH_8	$\{CH_{10}, BS\}$	CH_{10}
CH_9	$\{CH_{10}, BS\}$	BS
CH_{10}	$\{BS\}$	BS

Example 1 Let the dimension of every agent is 10 as $\vartheta = \{CH_1, CH_2, .., CH_{10}\}$ and the BS. The pictorial view of the subnetwork is depicted in Fig. 5. Every agent is initialized by random number between 0 and 1 as [0.21, 0.43, 0.13, 0.81, 0.76, 0.67, 0.31, 0.41, 0.52, 0.58]. Afterwards, the next-hop is calculated using Eq. 4.9. The computed next-hop for every CH is shown in Table 3.

4.3.3 Derivation of Fitness Function

(a) **Energy**: It is the residual energy of CH selected as a next-hop toward the BS. If the residual energy of all the selected CHs is high, it will increase the network lifetime as well as route will sustain to longer duration. Therefore, our first objective is g_1 which can be maximized as follows:

Objective 1

$$g_1 = \sum_{i=1}^{m} E_{CH_i} \tag{4.10}$$

(b) **Distance**: It is the distance from the CH to the next-hop (another CH) and from there to the BS. If this distance is minimum, it will expense less amount of energy in transmission and reduces the energy consumption in the routing path. Therefore, the second objective is g_2 which can be maximized as follows:

Objective 2

$$g_2 = \frac{1}{\sum_{i=1}^{m} \text{dis}(\text{CH}_i, \text{NH}) + \text{dis}(\text{NH}, \text{BS})} \tag{4.11}$$

All of the above-mentioned objectives are not strongly conflicting to each other. Therefore, weighted sum approach is applied and multiple objectives are converted into the single objective function as shown in Eq. 4.12. Here, Υ_1 and Υ_2 are the weighted value assigned to each objective.

Note: All the objectives are normalized before applying weighted sum approach.

$$\text{Maximize Fitness} = \Upsilon_1 \times g_1 + \Upsilon_2 \times g_2,$$

$$\text{where} \sum_{i=1}^{2} \Upsilon_i = 1, \Upsilon \epsilon (0,1); \tag{4.12}$$

4.3.4 Velocity and Position Update

In each iteration, velocity and position of an agent is updated using Eqs. (8 and 9). Every position is represented by real number between 0 and 1. Therefore, newly updated position should lie between 0 and 1. Due to the algebraic calculation, it may cause newly generated position is negative or greater than 1. In that case, a random number is considered between 0 and 1 to replace the corresponding position.

4.3.5 Pseudo Code for Routing Algorithm

Algorithm 2: GSA-Based Routing

Inputs: (1) Set of CH nodes $CH = (CH_1, CH_2, CH_3, \ldots, CH_m)$.
      (2) Predefined number of agents N_p.
      (3) Energy (CH_i), and Distance (CH_i).
      (4) Dimension of agent m.
      (5) Kbest=N_p.
Result: Near optimal routing path for each CH $\zeta = (\zeta_1, \zeta_2, \ldots, \zeta_m)$.

Step 1: Initialize the agent $A_i, \forall_i, 1 \le i \le N_p$.
Step 2:
while *(i! = N_p)* **do**
  | Calculate $Fitness(A_i)$ ▷ using Eq. 4.12
end
 Step 3:
while *(i! = N_p)* **do**
  | $best = maximum(Fitness(A_i))$
  | $worst = minimum(Fitness(A_i))$
end
Step 4:
while *(i! = N_p)* **do**
  | Calculate $mass(A_i)$ ▷ using Eq. 3.4
end
Step 5:
while *(s! = T)* **do**
  while *(i! = N_p)* **do**
    while *(j! = Kbest)* **do**
      5.1 Calculate the force of Agent A_i ▷ using Eq. 3.6
      5.2 Calculate the acceleration of Agent A_i ▷ using Eq. 3.7
      5.3 update the position of CH_i ▷ using Eqs. (3.8 & 3.9)
      5.4 update the Kbest value   ▷ using Eqs. (3.10 & 3.11)
    end
  end
end
Step 6: Calculate the route for each CH using maximum(fitness(A_i))

4.3.6 Illustrative Example of Route Selection Process Using GSA

Let the first position of an active agent is 0.5 with mass = 0.81, the first position of a passive agent is 0.36 with mass = 0.3, r (random number) is 0.5, R is 5 and T (maximum number of iterations) is 100.

The first position of ith agent updation is shown in Table 4. After that, map the position with next-hop using Eq. 4.9. In this way, the routing path between each CH and BS is found. Due to the algebraic calculation, if the new generated position may not satisfy the range, i.e., negative or greater than 1. In that case, the random number generated between 0 and 1 is taken to replace the corresponding position.

Table 4 Procedure for updating i^{th} agent in routing

Steps to calculate the routing path are as follows:
Step 1: Now calculate the gravitational constant $G(t)$, it reduces with time to control the search accuracy
$G(t) = G_0 * e^{-\alpha * t/T}$, where $G_0 = 100, \alpha = 20, t = 1$
$G(t) = 100 * e^{-20*1/100} = 83.33$
Step 2: Now, calculate the force between the active and passive agents using Eq. 6
Force $= \frac{G*M_a*M_b}{R+\varepsilon} * (x2 - x1) = (83.33 * 0.3 * 0.81) * (0.36 - 0.5)/(5) = -0.049$
Similarly, calculate force of each position of each agent
Step 3: Now, calculate the acceleration for first position of active agent a_1^1
$\text{Force}_1^{(1)} = \sum\limits_{j=1, j \neq 1}^{N_p} \text{rand} * \text{Force}_{ij}^1 = 0.26$
$a_1^1 = \frac{\text{Force}_1^{(1)}}{M_{ii}} = \frac{0.26}{0.81} = 0.320$
Step 4: Now, update the velocity and position of first position of an agent using Eqs. (8 and 9), respectively
$V_1^1 = r * V_1^1 + a_1^1 = 0.5 * 0 + 0.320 = 0.320$
$A_1^1(x) = A_1^1 + V_1^1 = 0.5 + 0.320 = 0.820$
Step 5: The above calculation updates velocity and first position. Therefore, first position of GSA (0.5) is updated to new position (0.820). In the similar way, every position of every agent is updated

Repeat the process iteratively until a maximum number of iteration is reached. After finding the near optimal routing path for each CH, start the communication process.

5 Simulation Result and Performance Evaluation

To evaluate/check the performance of proposed algorithms, extensive simulation is carried out using Matlab and Java programming. The parameters taken as input for WSN and GSA are shown in Tables 5 and 6. The performance tested in three different scenarios through variation in BS position, i.e., center (100×100), corner (200×200) and outside (250×250) the network. To show the performance of the proposed algorithm, we have executed some existing routing algorithms, such as DHCR proposed by Maryam and Reza [31], Hybrid Routing (HF) proposed by Abdulla et al. [29] and EADC proposed by Yu et al. [30]. All of the above algorithms executed 30 times and average results are taken for the comparison. To show the full potential of GSA (GSA-CHSR), we have also executed well-established techniques GA and PSO on same problem and platform with same fitness function, which termed as (GA-Routing) and (PSO-Routing).

Table 5 Parameters for WSN

Parameter	Value
Target area	$200 \times 200 \ m^2$
Base station location	Sin kX = (100–250), Sin kY = (100–250)
Number of sensor nodes	200–800
Number of CHs	20–60
Energy of CHs	2 J
d_o	30 m
d_{max}	100 m
E_{amp}	0.0013 pJ/bit/m^2
E_{fs}	10 pJ/bit/m^2
E_{elec}	50 nJ/bit
Packet size	4000 bits
Message size	500 bits

Table 6 Parameters for GSA

Parameter	Value
Number of agents (N_p)	20
K_{best}	15
V_{max}	200
L	1
Number of iteration	100
α_1	0.4
α_2	0.3
α_3	0.3
Υ_1	0.6
Υ_2	0.4

5.1 Performance Metric

To compare the performance of the proposed algorithm with existing algorithms, five test metrics are used namely, residual energy, number of alive nodes, network lifetime, data packets received at the BS, and convergence rate.

- **Residual energy**: In WSNs, every sensor node dissipates its energy in information transmission process which includes transmission, receiving, and aggregation. After dissipation, the leftover energy of a node is called residual energy. Network performance can be enhanced by increasing the residual energy of sensor nodes.

- **Number of alive nodes**: It is the number of nodes is left in working condition out of all in the network. So, it can be expressed in different ways such as half nodes are alive or some percentage of nodes are alive in the network. As the number of alive nodes increases, network performance increases.

- **Network lifetime**: It can be expressed in different ways. In this study, we have considered network lifetime as first node depth (FND). If network lifetime increases, it will increase the network performance.
- **Data packets received at BS**: It is the number of data packets received by the BS. It is directly proportional to both number of alive nodes and residual energy of the network. If network lifetime or residual energy increases, then it will also increase the number of data packets received at the BS.
- **Convergence rate**: It is expressed in terms of number of iterations taken by the algorithm to reach at the global optimal solution.

5.1.1 Residual Energy

The performance of the proposed and existing routing protocols in terms of residual energy over 200 sensor nodes is shown in Fig. 6. Hybrid routing (HF) is the combination of hierarchical and flat routing. Outside the hot-spot, it performs hierarchical routing and flat routing inside. In the NH selection process, only distance is taken as a decision parameter. But, lacking in residual energy consideration. Due to that reason sensors dissipate energy rapidly. Therefore, the performance of Hybrid routing is lowest. EADC is the hierarchical approach, it selects NH based on the residual energy and node degree. Therefore, it reduces the energy consumption in the network. In DHCR, CH selection is based on the domestic-Metric and peripheral-Metric. In domestic-Metric, residual energy is taken as a decision parameter to select the NH and peripheral-Metric takes distance to the BS and node degree. In performance, DHCR outperforms EADC and Hybrid routing. In GSA-CHSR, CH selection is based on the fitness function that considers distance as one of the objectives. Moreover, another fitness function also considers transmission distance from CH to NH and from there to the BS as one of the objectives in the routing process. Therefore, it reduces the energy consumption of each CH and performs better than comparatives.

In Fig. 6a–c, it is observed that HF and EADC dissipate energy rapidly, when BS was placed at corner (200 × 200) or outside (250 × 250) the target area. However, DHCR and GSA-CHSR still performs better. When the same problem is tackled by other optimization techniques, i.e., GA (GA-Routing) and PSO (PSO-Routing). It was found that GSA-CHSR performs better than both of them.

5.1.2 Number of Alive Nodes

The performance comparison of GSA-CHSR and its comparatives in terms of number of alive nodes over 200 sensors is shown in Fig. 7. In Fig. 7a, GSA-CHSR outperformed Hybrid routing, EADC, DHCR, PSO-Routing, and GA-Routing. In

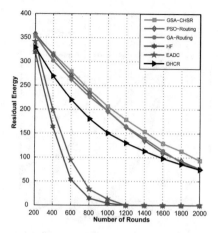

(a) Base Station at Center (100x100)

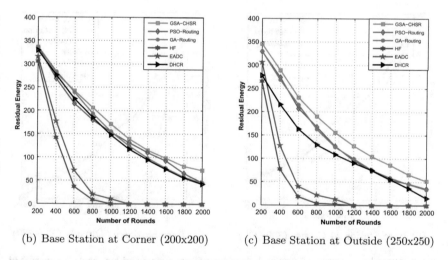

(b) Base Station at Corner (200x200)   (c) Base Station at Outside (250x250)

Fig. 6 Comparative results of proposed algorithm with existing algorithms as well as results of GA and PSO when applied on same fitness functions w.r.t residual energy

Fig. 7b, c, it is observed that the number of alive nodes decreases rapidly in HF, EADC, and DHCR, while slowly in GSA-CHSR. The reason behind the better performance of GSA-CHSR is that the fitness function takes care of residual energy as one of the objectives in CH selection and routing as well. It means energy consumption is less for low energy nodes and more for high energy nodes. In addition, it also balances the load on every CH. Therefore, number of alive nodes increases in the network and routing path sustain to the longer duration. When same fitness function was tested with other techniques, i.e., GA and PSO, it was found results of GSA-CHSR was better in most of the cases.

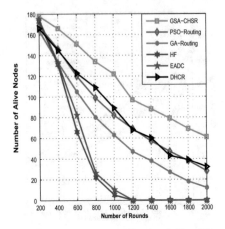

(a) Base Station at Center (100x100)

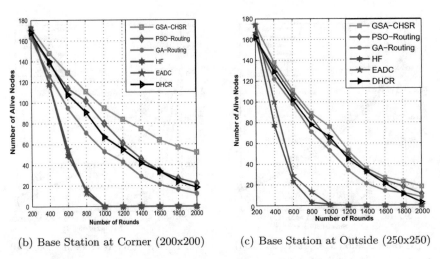

(b) Base Station at Corner (200x200)   (c) Base Station at Outside (250x250)

Fig. 7 Comparative results of proposed algorithm with existing algorithms as well as results of GA and PSO when applied on same fitness functions w.r.t number of alive nodes

5.1.3 Network Lifetime

The performance of the proposed algorithm is tested rigorously with varying number of sensor nodes from 200 to 800 and CHs from 60 to 90. In both the Fig. 8a, b, it is observed that the proposed algorithm outperforms the existing algorithms. The reason behind that the derived fitness function for CH selection considers residual energy. It means sensor nodes selected as CHs with high residual energy which reduces the probability of CH die quickly and prolong the network lifetime. Next, the residual energy is also considered for assigning sensor nodes to a

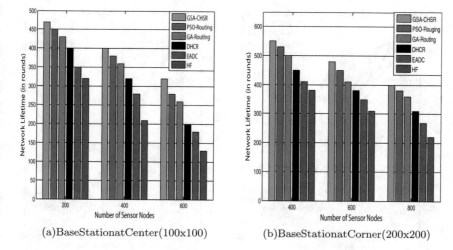

(a)BaseStationatCenter(100x100)   (b)BaseStationatCorner(200x200)

Fig. 8 Comparative results of proposed algorithm with existing algorithms as well as results of GA (GA-Routing) and PSO (PSO-Routing) when applied on same fitness functions w.r.t network lifetime

CH. Finally, the residual energy is considered for NH selection in the routing process.

5.1.4 Data Packets Received at the BS

The performance tested in terms of number of data packets received at the BS over 200 sensor nodes is shown in Fig. 9. As per the previous discussion in Sects. (5.1.2) and (5.1.1), GSA-CHSR performs better than HF, EADC, DHCR, PSO-Routing,

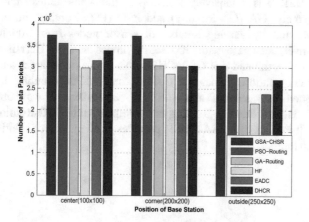

Fig. 9 Comparison of GSA-CHSR with existing algorithms

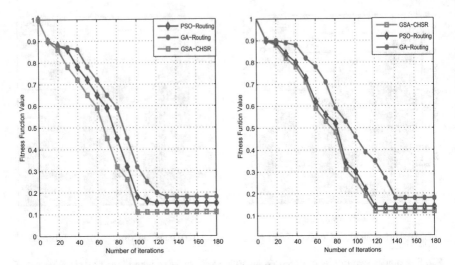

Fig. 10 Comparative results of GSA-CHSR with GA (GA-Routing) and PSO (PSO-Routing) when applied on same fitness functions w.r.t convergence rate

and GA-Routing in terms of residual energy and number of alive nodes. The number of data packets received at the BS is directly determined by residual energy and number of alive nodes as mentioned in Sect. 5.1. Therefore, number of data packets received at the BS more in GSA-CHSR compared to existing routing methods. It is proved that proposed routing enhance the aforementioned purpose manifold.

5.1.5 Convergence Rate

The GSA-CHSR was extensively tested with the well-establish nature inspired algorithm namely, GA (GA-Routing) and PSO (PSO-Routing) in terms of convergence rate and by varying number of sensor nodes from 300 to 400. The obtained convergence rate with 300 sensor nodes and 30 CHs is depicted in Fig. 10a. It was estimated that GSA-CHSR was quickly converges and reaches to nearer optimum solution compared to GA-Routing and PSO-Routing. In the second scenario, when 400 sensors and 40 CHs was taken, it was observed that GSA-CHSR outperforms GA-Routing and competitive to PSO-Routing as shown in Fig. 10b. It implies that GSA-CHSR provides good quality of solution with less number of iterations.

6 Conclusion

In this study, an effort has been made to establish GSA as an efficient algorithm for solving WSN-related optimization problems, i.e., cluster head selection and routing. A novel objective function, based on the energy, distance, and CH balancing factor, was derived for cluster head selection and a GSA-based routing algorithm was also devised with residual energy and distance(s) as its fitness function parameters. GSA-CHSR was rigorously tested with recent existing algorithms, such as Hybrid Routing, EADC, and DHCR. The obtained results of GSA-CHSR were better in most of the cases on various scenarios compared to existing ones. To show the full potential of GSA, both the problems were also solved by other nature inspired optimization techniques, such as GA (GA-Routing) and PSO (PSO-Routing) with same fitness functions. It was found that in most of the cases GSA-CHSR was superior to PSO-Routing and GA-Routing in terms of performance metrics. Moreover, it was competitive to PSO-Routing and better than GA-Routing in terms of convergence rate.

References

1. Singh AK, Purohit N, Varma S (2013) Fuzzy logic based clustering in wireless sensor networks: a survey. Int J Electron 126–141
2. Jian JC, Ren WC, Min X, Lun TX (2010) Energy-balanced unequal clustering protocol for wireless sensor networks. J China Univ Posts Telecommun 17(4):94–99
3. Mao S, Zhao C, Zhou Z, Ye Y (2014) An improved fuzzy unequal clustering algorithm for wireless sensor network. Mobile Netw Appl 206–214
4. Kumar D, Aseri TC, Patel RB (2009) Energy efficient heterogeneous clustered scheme for wireless sensor networks. Comput Commun 32:662–667
5. Li C, Ye M, Chen G, Wu J (2008) An energy-efficient unequal clustering mechanism for wireless sensor networks. IEEE Int Conference Mobile Ad-hoc Sensor Syst 1–8
6. Ran G, Zhang H, Gong S (2010) Improving on LEACH protocol of wireless sensor networks using fuzzy logic. J Inf Comput Sci 7(3):767–775
7. Li H, Liu Y, Chen W, Jia W, Li B, Xiong J (2013) COCA: constructing optimal clustering architecture to maximize sensor network lifetime. Comput Commun 36(3):256–268
8. Bagci H, Yazici A (2010) An energy aware fuzzy unequal clustering algorithm for wireless sensor networks. In: Proceedings of the IEEE international conference on fuzzy system, pp 1–8
9. Liu AF, You WX, Gang CZ, Hua GW (2010) Research on the energy hole problem based on unequal cluster-radius for wireless sensor networks. Comput Commun 33(3):302–321
10. Taheri H, Neamatollahi P, Younis OM, Naghibzadeh S, Yaghmaee MH (2012) An energy-aware distributed clustering protocol in wireless sensor networks using fuzzy logic. Ad-hoc, Netw 10(7):1469–1481
11. Agarwal PK, Procopiuc CM (2002) Exact and approximation algorithms for clustering. Algorithmica 33(2):201–226
12. Ferng HW, Tendean R, Kurniawan A (2012) Energy-efficient routing protocol for wireless sensor networks with static clustering and dynamic structure. Wirel Pers Commun 65(2): 347–367
13. Awwad SAB, Ng CK, Noordin NK, Rasid MFA (2011) Cluster based routing protocol for mobile nodes in wireless sensor network. Wirel Pers Commun 61(2):251–281

14. Dorigo M, Birattari M, Stutzle T (2006) Ant colony optimization. IEEE Comput Intell Mag 1(4):28–39
15. Rashedi E, Nezamabadi-pour H, Saryazdi S (2009) GSA: a gravitational search algorithm. Inf Sci 179:2232–2248
16. Heinzelman WB, Chandrakasan AP, Balakrishnan H (2002) An application specific protocol architecture for wireless microsensor networks. IEEE Transac Wireless Commun 1:660–670
17. Younis O, Fahmy S (2004) A hybrid energy-efficient, distribution clustering approach for ad-hoc sensor networks. IEEE Transac on MC 366–379
18. Wang A, Yang D, Sun D (2012) A clustering algorithm based on energy information and cluster heads expectation for wireless sensor networks. Comput Electr Eng 38:662–671
19. Heinzelman WR, Chandrakasan A, Balakrishnan H (2000) Energy-efficient communication protocol for wireless microsensor networks. In: Proceedings of the 33rd Hawaii international conference on system sciences, pp 1–10
20. Chang JY, Ju PH (2012) An efficient cluster-based power saving scheme for wireless sensor networks. EURASIP J Wireless Commun Netw 172:1–10
21. Yang J, Ju PH (2014) An energy-saving routing architecture with a uniform clustering algorithm for wireless sensor networks. Future Gener Comput Syst 36:128–140
22. Bagci H, Yazici A (2013) An energy aware fuzzy approach to unequal clustering in wireless sensor networks. Appl Soft Comput 13(4):1741–1749
23. Lee JS, Cheng WL (2012) Fuzzy-Logic-Based clustering approach for wireless sensor networks using energy prediction. IEEE Sens J 12(9):2891–2897
24. Kumar SS, Kumar MN, Sheeba VS (2011) Fuzzy logic based energy efficient hierarchical clustering in wireless sensor networks. Int J Res Rev Wirel Sens Netw 53–57
25. Banerjee S, Khuller S (2001) A clustering scheme for hierarchical control in multi-hop wireless networks. In: Proceedings of IEEE INFOCOM
26. Gerla M, Kwon TJ, Pei G (2000) On demand routing in large ad hoc wireless networks with passive clustering. In: Proceeding of WCNC
27. Senouci MR, Mellouk A, Senouci H, Aissani A (2012) Performance evaluation of network lifetime spatial-temporal distribution for WSN routing protocols. J Netw Comput Appl 35:1317–1328
28. Lai Wk, Fan CS, Lin LY (2012) Arranging cluster sizes and transmission ranges for wireless sensor networks. Inform Sci 183(1):117–131
29. Abdulla AEAA, Nishiyama H, Kato N (2012) Extending the lifetime of wireless sensor networks: a hybrid routing algorithm. Comput Commun 35:1056–1063
30. Yu J, Qi Y, Wang G, Gu X (2012) A cluster-based routing protocol for wireless sensor networks with nonuniform node distribution. Int J Electron Commun 66:54–61
31. Maryam S, Reza NH (2015) A decentralized energy efficient hierarchical cluster-based routing algorithm for wireless sensor networks. Int J Electron Comm (AEAœ)
32. Song M, Cheng-lin Z (2011) Unequal clustering algorithm for WSN based on fuzzy logic and improved ACO. J Chin Univ Posts Telecommun 18:89–97
33. Bhari A, Wazed S, Jaekal A, Bandyopadhyay S (2009) A genetic algorithm based approach for energy efficient routing in two-tiered sensor networks. Ad Hoc Netw 7:665–676
34. Elhabyan RSY, Yagoub MCE (2015) Two-tier particle swarm optimization protocol for clustering and routing in wireless sensor network. J Netw Comput Appl

Utilizing Genetic-Based Heuristic Approach to Optimize QOS in Networks

Sherin Zafar

1 Introduction

With the growth-based networks using wireless framework, the performance of these networks leads to the reliability of Internet which in turn largely depends on how the routing protocols operate [1]. The operation of these routing protocols is dependent upon the network technology, its configuration parameters focusing upon the traffic load on the routers and its links. Specifying the concept of wireless systems routing becomes a highly challenging task due to the mobility of these types of systems. The traditional routing protocols utilized in the current networks are developed to exploit dynamic optimization by finding optimized paths based on shortest distance, minimal bandwidth and delay and limited power and capability of links, etc. Number of topology changes frequent in nature occur in wireless-based cloud systems due to mobility of wireless nodes and their energy conservation therefore, dynamic optimization problem (DOP) for route discovery becomes quite important. For finding solutions of various routing issues, traditional methodologies utilizing deterministic algorithms like shortest path tree (SPT) and search heuristics based neoteric approaches like genetic algorithm, ant colony optimization, simulated annealing, etc., are explored. Deterministic systems, on one hand, construct only on one route tree for a given route discovery request whereas meta-heuristic based algorithms search number of route trees and from them select optimized one as the final route, as these algorithms have polynomial time complexity, hence providing enhanced Quality of Service (QOS) based solutions for cloud-based wireless networks. When compared with deterministic systems (Previous Dijkstra's Algorithm), GA-based meta-heuristic approach works on a population of possible solutions rather than on a single solution. The individual chromosomes (initial

S. Zafar (✉)
Department of Computer Science & Engineering, School of Engineering Sciences &
Technology, Jamia Hamdard (Deemed to be University), New Delhi, India
e-mail: zafarsherin@gmail.com

© Springer Nature Singapore Pte Ltd. 2017
R. Ali and M. M. S. Beg (eds.), *Applications of Soft Computing for the Web*,
https://doi.org/10.1007/978-981-10-7098-3_14

population) of the GA-based approach undergo selection operation on the basis of their fitness value followed by the natural genetic-based crossover operation for offspring generation by randomly exchanging the individual genetic data focusing on the fact that GA is stochastic in nature rather than deterministic [2–6]. This chapter exploits GA-based methodology of soft computing for optimization of various QOS parameters like packet delivery ratio (PDR), packet drop rate (pdr), end-to-end delay (EED) and hop count (HC) for wireless-based cloud networks by selection of optimal or fitness route from a given source to a given destination from a route set of different connectivity qualities. Upcoming sections of this chapter will highlight on utilizing soft computing based GA approach for optimization of QOS in wireless-based cloud networks by validation through simulated results.

2 Importance of Genetic Algorithm

Natural evolution is the main source of inspiration for genetic algorithms for becoming a heuristic and optimization methodology proving to be applied successfully for various complex and significant real-world problems. This section of the chapter focuses on utilizing the success of GA-based approach for improving the quality of service (QOS) parameters of traditional routing approaches (e.g. Dijkstra's) being used in the current routing protocols of wireless-based networks or cloud networks. GA has been a great interest for various mathematicians, immunologists, etc. Before discussing how GA has enhanced QOS parameters of wireless and cloud networks (discussed in next section), we will first focus on the GA construction and its theoretical specific strands speculatively. Darwinian principle of evolution is the main motivation for GA-based heuristic and solution search heuristic and solution search based optimization methodology. For evolving for solutions to various problems GA utilizes evolutionary processes which are highly abstracted in version. Populations of GA operation are the chromosomes which are binary strings. Fitness function of GA-based approach is a real number in the solution representation of chromosomes specifying how good solution is for a particular problem [7–16].

The successor population or the next generation of GA is produced through fitness selection and recombination of the randomly generated population of chromosomes. Recombination process selects the genetic material of parent chromosomes which undergoes recombination to form child chromosomes and continues further inheritor population. Process iteration evolves a successive generation sequence leading to increase in the chromosomes average fitness until reaching a stopping criterion. Hence, GA evolved as a best (optimized) solution for a given problem. For computationally intractable problems, Holland [17] proposed GA for finding good solutions. The theoretical concepts of GA and its straight forward implementation were proved by Schema Theorem written by Goldberg [18] and its block hypothesis [17]. These results led to the growth of GA in solutions of practical problems of sciences, industry and engineering also discussed in this chapter for QOS-based route optimization in cloud-based wireless networks. GA-based approach is quite dynamic in nature using methodologies not utilized by traditional approaches, hence its success leads to the growth of other intelligent meta-heuristic based approaches like neural network (NN), ant colony optimization (ACO), particle swarm optimization (PSO), artificial intelligence (AI), etc. [19].

2.1 Structure of Genetic Algorithm

Implementation of GA is not very complex as it is constructed from distinct reusable components having trivial adaptation. The components of GA which form its basic structure include encoding of chromosomes, development of fitness function, selection of population based on fitness function, recombination and scheme evolution being the last operations. Figure 1 shows the evolution flow of genetic algorithm.

(i) Chromosome Encoding

Chromosomes in a GA for a particular problem are the representation of strings represented as {A, C, G, T} which is a biological abstraction of DNA chromosome. Locus is the particular position in a chromosome called a gene, and the letter at the point of locus is called as allele value or allele. Allele values are the bit strings of chromosomes that are represented as character alphabets {0.1}. GA operates on bit strings of length around 100 for a solution space of 2100–1030 individuals.

(ii) Fitness

Fitness function for a particular problem specifies computation by evaluating chromosomes of genetic algorithm.

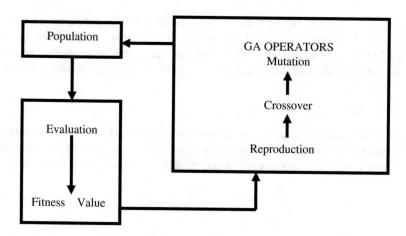

Fig. 1 Evolution flow of genetic algorithm

(iii) Selection

Selection process of GA utilizes fitness function for the evolution of the genetic chromosomes by applying a selective pressure. On the basis of their fitness, chromosomes are selected for recombination where better is the fitness of chromosomes greater is the chances of selection. Roulette wheel, random stochastic selection, tournament selection and truncation selection are generally utilized as traditional selection methods.

(iv) Recombination

Members of a successor population are formed through recombination process by selected chromosomes of a given population. This is done by simulating the mixed material occurring at the tie of organisms' reproduction. Crossover and mutation are the two basic components of the recombination process where crossover produces one/two child chromosomes by mixing genetic material of the two parent chromosomes being selected.

The given example below represents the crossover operation done over a 10-bit string.

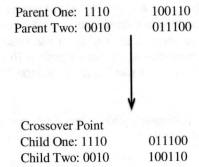

Parent One: 1110   100110
Parent Two: 0010   011100

Crossover Point
Child One: 1110   011100
Child Two: 0010   100110

Mutation operation of GA flips one/more allele values by acting on the various individual chromosomes. When mutation rate $\geq$ the random value, then the flipping of allele values takes place as 0–1 or vice versa. For better results and overcoming the problem of local/global minima, mutation rates are chosen quite small like 0.001, etc.

(v) Evolution

Many evolutionary schemes are based upon how much the initial chromosomes are unchanged with respect to the inheritor population. Complete replacement and replacement with elitism are some of the methods utilized for the evolution.

2.2 Design of GA

The nature of the problem is the most important criterion for the design and encoding of GA. After the encoding methodology is being selected, next decision is about the type of fitness function, size of the population, respective rates of crossover and mutation operators, application of the evolutionary scheme and also the specific and appropriate start and stop conditions. The design of GA basically depends upon experience of combination, modelling which is problem specific and experimentation based upon various evolution schemes [20–25]. A typical and classical GA design based upon complete replacement using standard operators of GA includes

(1) Random generation of a population comprising of P chromosomes.
(2) From the source population of initial chromosomes (C), fitness function $F(C)$ is calculated.
(3) A vacuous successor population is created and the following steps are repeated till creation of P chromosomes takes place.

    (a) Two chromosomes $c1$ and $c2$ are selected based upon the proportional fitness function from the initial source population.
    (b) Utilizing crossover rate pc one point crossover is applied to obtain c as the child chromosome.
    (c) A consistent rate of mutation pm is applied to produce c'.
    (d) c' is being added to the inheritor population.

(4) Source population is then replaced with the population of successor.
(5) Step 2 is being returned, not meeting the stopping criterion.

2.3 Approaches of GA

The basic two goals for justifying theory of GA include explaining the classes of problem to which GA is particularly suitable and if suitable, why. Next includes the optimal design and implementation of GA through various techniques and approaches.

(i) Schema and the Building Blocks: The central result of GA theory is based upon the Schema Theorem [17]. A set of chromosomes containing the pattern specifies the schema. Schema is referred as the string symbols from the alphabet {0, 1, *}. For chromosomes of length 10 bit, schema specifies using the string

```
01***100**
```

Chromosomes belonging to the schema include

Chromosomes that do not belong to the schema, not matching the pattern, include

All the chromosomes that belong to the schema are **********.

For a schema X, the length lX is specified as antithesis of the allele positions of the first bits and the last bits that are being defined of X. The numbers of bits being defined specify order of X. For example, the schema 01***100** has defined bits that has the order 5. The position of last defined bit is 8 and that of first is at position 1. So, the length of schema is $8 - 1 = 7$.

Theorem is defined by taking X as schema and chromosomes as mX that belong to X that is present in the population I of the GA being evolved. Then the number of is chromosomes expected to belong to X in the population $i + 1$ is denoted by $mX (i + 1)$, is represented by the formula

$$mX(i+1) = FX(i)mX(i)[1 - pc\, lX/l - 1][(1 - pm)OH] \qquad (1)$$

where

FX (i) is the relative fitness of the chromosomes that belong to X that is being divided by the average fitness of the population of chromosomes.

pc is the probability of crossover operation.

pm is the probability of nutation operation.

(ii) The Simple GA (SGA): SGA provides a dynamic mathematical framework for exploring fundamental properties of GA. It answers to the various questions expected

(a) Locus Ω in SGA at time t?
(b) Convergence of SGA at a global optimum based on what conditions?

(c) Design choice of GA based on conditions of point (b).
(d) What will be the mutation and selection operators?
(e) What will be the selection schemes?
(f) How the evolution points correspond to local/global minima.

(iii) Statistical mechanics: [26] developed and utilized the theory of statistical mechanics for GA evolution having a fitness function (FF) based on Ising Model that has states of low energy of a spin glass through which investigation of dynamics for GA is done.

(iv) GA is utilized in model fitting and optimal control in various applications, e.g. cancer chemotherapy optimization by GA.

3 Genetic-Based Heuristic Approach to Optimize QOS in Networks

Depending upon the various linguistic rules, network conditions currently and the policy, GA performs evaluation of the paths; output of the routing algorithm is resulted, after evaluating all the routes having good enough route values depending upon the evaluation function [27–30]. After performance evaluation of the values, the next loops are being selected from the various neighbouring nodes. The selection of the routing by the GA-based heuristic approach is done by optimizing QOS parameters like packet delivery ratio (PDR), hop count (HC), etc. The flowchart of genetic-based heuristic approach is depicted in Fig. 2. Given below are the specific parameters required for the GA-based heuristic approach

1. Genetic-based representation specifying node ID's of the path of the source node through gene's first position and last position gene represents the destination node.
2. Population initialization (PI) is the randomly generated routing path from the initial population chromosomes.
3. Fitness function (FF) is represented by function *fval* containing the object function including the arguments as initial route x, a matrix containing a proactive/reactive routing table, start location and end location. Route cost is initiallized as 0 and further values of route cost are obtained on the basis of steps given below:

   (a) Value of $x1$ is generated by subtracting 1 from size matrix and multiplying by x.
   (b) Next round $x1$ is generated as $x1 = (A, B)$, where A = absolute value of $x1$ rounded and mod operator applied, B = increment of size matrix by 1.
   (c) A condition is applied to generate the route cost where comparison (==) is done between $x1$ and x which is further incremented by 2.

(d) A for loop is followed, then route cost is generated as $x1$ matrix ith value, $x1$ matrix $(i + 1)$ value added to initial route cost which is considered f value of route cost. Else the final value of route cost is size of matrix multiplied by 100 and added to 100.

4. Selection operator selects higher quality chromosomes based on FF to get the next generation.
5. Crossover and mutation are utilized to exchange partial chromosomes to overcome local/global minima issue.
6. GA-based approach is utilized as a solution for finding the shortest path from source (S) to destination (D) dynamically in a peer-to-peer cloud network. GA considers a peer-to-peer cloud network having an upper delay bound with a

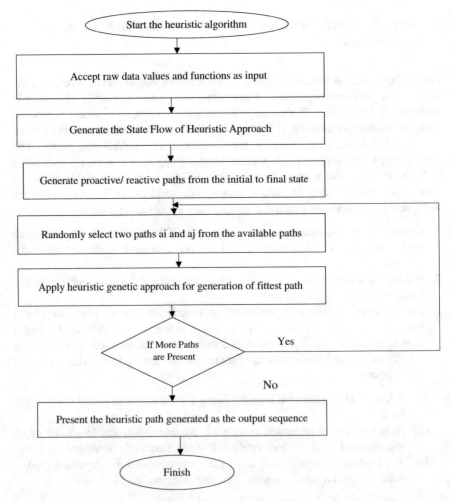

Fig. 2 Flowchart for genetic-based heuristic approach to optimize QOS in networks

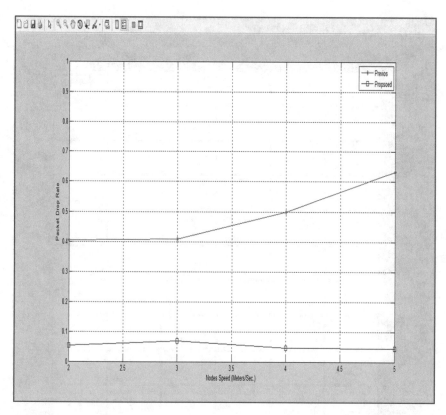

Fig. 3 Genetic algorithm output showing the effectiveness of optimized QOS parameter packet drop rate with node speed in the X axis

    source node and destination node to get an optimized least cost path in a topological area.
7. The QOS parameters like packet delivery ratio, packet drop rate, end-to-end delay and hop count are taken into consideration for performance analysis.

The simulation parameters taken into analysis of the meta-heuristic based genetic approach to optimize QOS in cloud networks are simulation time, network length and width, nodes of peer-to-peer cloud network, transmission range, node speed, node data rate, node traffic and speed and angle variation factor [31–33]. Figures 3, 4, 5, 6 and 7 specify the validation of the proposed approach through a MATLAB simulator.

Figure 3 depicts the simulation of previous (Dijkstra's) approach and proposed (Genetic-based heuristic) approach through varying node speed, validating the

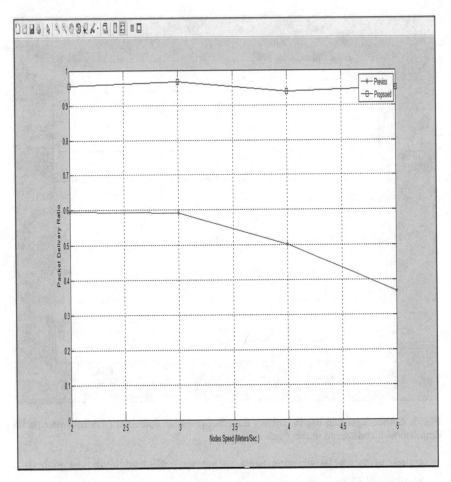

Fig. 4 Genetic algorithm output showing the effectiveness of optimized QOS parameter packet delivery rate with node speed in the X axis

proposed approach through blue line, specifying how through the application of heuristic approach, the packet drop rate has decreased significantly.

Figure 4 depicts the simulation of previous (Dijkstra's) approach and proposed (Genetic-based heuristic) approach through varying node speed, validating the proposed approach through blue line, specifying how through the application of heuristic approach, the packet delivery ratio has increased significantly.

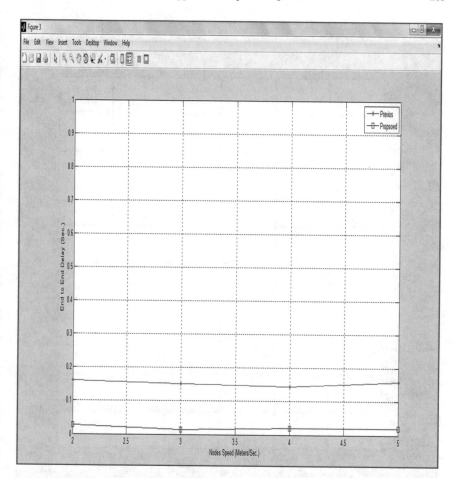

Fig. 5 Genetic algorithm output showing the effectiveness of optimized QOS parameter end-to-end delay with node speed in the X axis

Figure 5 depicts the simulation of previous (Dijkstra's) approach and proposed (Genetic-based heuristic) approach through varying node speed, validating the proposed approach through blue line, specifying how through the application of heuristic approach, the end-to-end delay has decreased significantly.

Figure 6 depicts the simulation of previous (Dijkstra's) approach and proposed (Genetic-based heuristic) approach through varying node speed, validating the

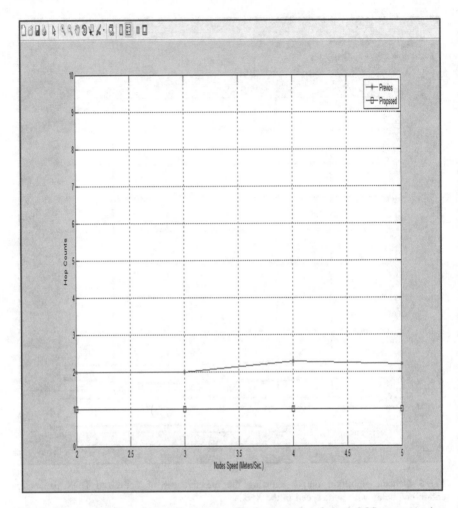

Fig. 6 Genetic algorithm output showing the effectiveness of optimized QOS parameter hop counts with node speed in the *X* axis

proposed approach through blue line, specifying how through the application of heuristic approach, the hop counts have decreased significantly.

Figure 7 is just a background illustration of working of genetic algorithm which validates the proposed approach.

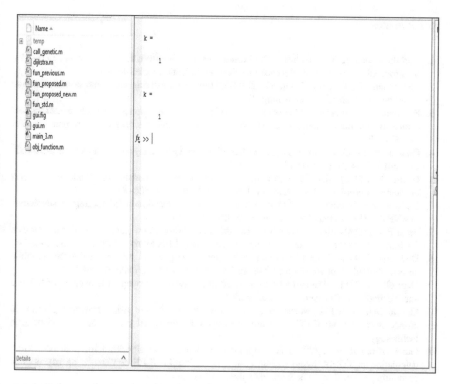

Fig. 7 Background working of genetic algorithm

4 Conclusions

Talking about real-world peer-to-peer wireless network, routing is considered to be the parameter that will meet the most stringent QOS performance parameters. Uncertainty that is being created through user mobility in peer-to-peer networks makes the seamless data transmission quite a challenging task. A number of conflicting issues are raised in these networks. In real-world scenario when a particular objective function is optimized, other dependent objective can be sacrificed. The results shown in Sect. 3 specify how the heuristic-based genetic approach has led to neoteric optimized results mainly exploited to finding fittest shortest path leading to optimized QOS performance parameters. The conventional shortest path selection approach like Dijkstra's (taken as previous approach in Sect. 3 simulation) was not able to provide better performance in terms of QOS optimization. Taking into consideration number of simulation parameters, the simulated, validated approach discussed in this chapter provides a flexible network by enhancing various QOS issues.

References

1. Abdullah J et al (2008) GA based QOS route selection algorithm for mobile ad-hoc networks. In: Proceedings of IEEE conference on telecommunication technologies
2. Begumhan TD, Turgut R, Than VL (2003) Optimizing clustering algorithm in mobile adhoc networks using simulated annealing. IEEE
3. Bellavista P, Corradi A, Gianneli C (2011) A unifying perspective on context-aware evaluation and management of heterogeneous wireless connectivity. IEEE Commun Surv Tut 13:337–357
4. Cizmar A, Papaj J, Dobos L (2012) Security and QOS integration model for MANET. Comput Inform 31:1025–1044
5. Cheng H (2010) Genetic algorithms with immigrants schemes for dynamic multicast problems in mobile ad-hoc networks. Eng Appl Artif Intell, 806–819
6. Defrawy KE, Tsudik G (2011) ALARM: anonymous location-aided routing in suspicious MANET's. IEEE Trans Mob Comput 10(9):1345–1358
7. Fessi BA, BenAbdullah S, Hamdi M, Boudriga N (2009) A new genetic algorithm approach for intrusion response system in computer networks. IEEE Symp Comp Commun, 342–347
8. De Rango F, Socievole A (2011) Meta-heuristics techniques and swarm intelligence in mobile ad-hoc networks. In: Book on mobile ad-hoc network and applications, vol 11
9. Gabrielle A (2011) Simulation of a secure ad-hoc network. Norwegian University of Science and Technology, Department of Telematics
10. Ghazal MA, Sayed A, Kelash H (2007) Routing optimization using genetic algorithm in ad-hoc networks. In: IEEE international symposium on signal processing and information technology
11. Gunasekaran R et al (2009) An improved parallel genetic algorithm for path bandwidth calculation in TDMA based mobile ad-hoc networks. In: IEEE conference on advances in computing, control and telecommunications technologies
12. Jagadeesan AT, Duraiswamy K (2010) Cryptographic key generation from multiple biometric modalities: Fusing minutiae with Iris feature. Int J Comput Appl 2(6)
13. Jyotika K, Baregar AJ (2013) Security using image processing. Int J Manag Info Technol 5(2)
14. Krishna BA, Radha S, Reddy KCK (2007) Data security in ad-hoc networks using randomization of cryptographic algorithms. J Appl Sci, 4007–4012
15. Nandi B, Barman S, Paul S (2010) Genetic algorithm based optimization of clustering in ad-hoc networks. Int J Comput Sci Inf Secur 7(1)
16. Sanderson S, Erbetta J (2000) Authentication for secure environments based on Iris scanning technology. IEEE Colloq Vis Biomet
17. Holland JH (1975) Adaptation in natural and artificial systems. The University of Michigan Press, Ann Arbor, MI
18. Goldberg DE (1989) Genetic algorithms in search, optimization and machine learning. Addison-Wesley, Reading, MA
19. Engelbrecht AP (2002) Computational intelligence an introduction. Wiley, New York
20. Sanzgiri K, Laflamme D, Dahill B, Levine BN (2005) Authenticated routing for ad-hoc network. IEEE J Sel Areas Commun
21. Shannon RE (1989) Introduction to the art and science of simulation. In: Proceedings of 30th conference on winter simulation
22. Zafar S, Soni MK, Beg MMS (2015) An optimized genetic stowed approach to potent QOS in MANET. Proc Comput Sci 62:410–418. https://doi.org/10.1016/j.proc.08.434. ISSN: 1877-0509

23. Zafar S, Soni MK (2014) Sustaining security in MANET: biometric stationed authentication protocol (BSAP) inculcating meta-heuristic genetic algorithm. Int J Mod Educ Comput Sci 9:28–35. Published Online September in MECS. http://www.mecs-press.org/,, https://doi.org/10.5815/ijmecs.2014.09.05. ISSN: 2075-0161(Print), ISSN: 2075-017X (Online) Impact Factor 0.669 (2015)
24. Zafar S, Soni MK, Beg MMS (2014) Sustaining security: encircling wavelet quartered extrication algorithm for crypt-biometric perception. In: International conference on data mining and intelligent computing (ICDMIC). IEEE Conference Publications, pp 1–6. https://doi.org/10.1109/ICDMIC.2014.6954263
25. Zafar S, Soni MK (2014) Trust based QOS protocol (TBQP) using meta-heuristic genetic algorithm for optimizing and securing MANET. IEEE Explore, pp 173–177. https://doi.org/10.1109/ICROIT.2014.6798315
26. Prügel-Bennett A, Shapiro JL (1994) An analysis of genetic algorithms using statistical mechanics. Phys Rev Lett 72(9):1305–1309
27. Zafar S, Soni MK (2014) A novel crypt-biometric perception algorithm to protract security in MANET. Int J Comput Netw Inf Secur 6(12):64–71. https://doi.org/10.5815/ijcnis.12.08. ISSN: 2074-9090 (Print). ISSN: 2074-9104 (Online), https://doi.org/10.5815/ijcnis. Published by: MECS Publisher, November 2014, Impact factor 0.726 (2015)
28. Umadevi V, Chezhian R, Khan ZU (2012) Security requirements in mobile ad-hoc networks. Int J Adv Res Comput Commun 1(2)
29. Yen YS et al (2008) A genetic algorithm for energy-efficient based multicast routing on MANET. J Comput Commun, 2632–2641
30. Zarza L, Pegueroles J, Soriano M (2007) Interpretation of binary strings as security protocols for their evolution by means of genetic algorithms
31. Sherin Z, Soni MK (2014) Secure routing in MANET through crypt-biometric technique. In: Proceedings of the 3rd international conference on frontiers of intelligent computing: theory and applications (FICTA), pp 713–720
32. Zafar S, Soni MK, Beg MMS (2015) QOS optimization in networks through meta-heuristic quartered genetic approach. ICSCTI, IEEE
33. Zafar S, Soni MK, Beg MMS (2016) Iris signature methodology for securing MANET. MR Int J Eng Technol 8(1)

Part X
Other Emerging Soft Computing Techniques & Applications

V-MFO: Variable Flight Mosquito Flying Optimization

Md Alauddin

1 Introduction

The world would be more stunning if all the activities had been set at their ideal values. In practice, the ideal values of the variables, parameters, and constants of a process are determined by optimization techniques. In fact, optimization is the search for the best possible solution(s) to a given problem while satisfying its constraints. It has been an active area of research for several decades. With a broad range of applications in engineering, science, management and industries, it has become a powerful problem-solving method. Eventually, it has developed as a mature field in the form of many branches such as linear, nonlinear, MINLP, convex, discrete, global, and dynamic optimization. Despite this development, real-world optimization problems are becoming increasingly complex. Thus, better optimization algorithms are always needed.

To solve complex problems, researchers are working on a new direction to learn and adopt natural intelligence in their algorithms, because nature is the penultimate example of design and perfection. As a result, several naturally inspired optimization techniques have been developed over the last three decades [1]. The similarity in all these algorithms is that they work with multiple points, known as population, instead of a single point like in the conventional optimization techniques [2]. These multiple points do not become trapped at local peaks of a multimodal function. Some of the outstanding meta-heuristic algorithms include genetic algorithm (GA) [3, 4], ant colony optimization (ACO) [5], particle swarm optimization (PSO) [6], and artificial bees colony (ABC) [7]. A few recently proposed meta-heuristic algorithms comprise cuckoo search [8], seven-spot ladybird optimization (SLO) [9], bacterial foraging algorithm (BFA) [10], flower pollination

Md. Alauddin (✉)
Department of Process Engineering, Memorial University of Newfoundland,
St. John's, Canada
e-mail: alauddinchem@gmail.com

© Springer Nature Singapore Pte Ltd. 2017
R. Ali and M. M. S. Beg (eds.), *Applications of Soft Computing for the Web*,
https://doi.org/10.1007/978-981-10-7098-3_15

optimization [11], lion optimization [12], monkey optimization [13], eagle optimization [14], wolf optimization [15], whale optimization [16], krill herd (KH) [17], Ions motion algorithm [18], and mosquito flying optimization (MFO) [19].

These meta-heuristic algorithms have been extensively applied in different fields, ranging from manufacturing to services, scheduling to transportation, health to sports, geology to astronomy, constrained to unconstrained, single objective to multiple objectives, static to dynamic, and continuous to discrete. However, an argument against these algorithms is that they do not perform equally well in all cases. For instance, an algorithm which performs better in a particular application may not be effective in another class of problems [20]. Nonetheless, all exhibit the basic characteristics of intensification and diversification. Parenthetically, intensification represents an idea of search around the current best solution, whereas diversification speaks for the exhaustive exploration over the entire search space. The success of an optimization algorithm depends on the balancing between these explorations and exploitations.

In this work, a new nature-inspired population-based method, namely the variable mosquito flying optimization (V-MFO), is proposed which mimics the behavior of mosquitoes to find holes in a mosquito net. It incorporates the variable flying constant and the precision movements of the proboscis instead of a constant flying, and sliding motion, which makes it different from the MFO algorithm. The paper is structured in the sections with a brief outline of the algorithm followed by experimentation with benchmark functions, and finally a comparison with the acknowledged algorithms.

2   The Variable Flight Mosquito Flying Optimization (V-MFO) Algorithm

Mosquitoes belong to bloodsucking insects characterizing the family of flies referred to as Culicidae [21]. They are very diverse in nature, with more than 3500 species worldwide. Anatomically, they have got slim bodies with three segments: a head, a thorax, and stomach. They possess an average lifespan of as short as a week to quite a few months long, depending on species, sex, and weather conditions. The mosquitoes go through four lifecycle stages, similar to other flies: egg, larva, pupa, and adult. They feed mainly on nectar, plant juices, and decaying plant materials. However, the female mosquitoes also suck blood for protein to produce eggs [21]. While sucking, they release saliva that causes an irritating rash, which can be a serious allergy or nuisance. As a matter of fact, they are carriers of several dangerous diseases, such as malaria, yellow fever, and dengue fever to name a few [22]. This makes them one of the deadliest creatures on earth. That is why human beings make use of eclectic preventives in the form of spray, creams, coils, and net.

A mosquito has diverse approaches of locating their prey, including chemical, visual, and heat sensors [23].

Mosquitoes hover around a net in search of holes or irregularities that may facilitate them to reach at their host as shown in Fig. 1. In order to get an entry point, they attempt flying as well as sliding motion, i.e., they fly to a point on the net and walk in any direction for a few steps as depicted in Fig. 1a. If they succeed in finding holes or irregularities, they get into the net and search for a suitable suction point on the body of the victim. The distinct movements of the mosquitoes over a net are described in Fig. 1b–d. Figure 1b depicts that a mosquito flying from point "A", reaches point "B" by a flying motion (solid line) and then continues to point "C" using a sliding motion (dotted line). The sliding motion could be in the same direction to the flying motion that may lead to point "C", or it may be in the opposite direction to facilitate reaching to point "C'". However, this is generally in the same direction. The difference between flying and sliding motion is that, in flying the distance covered is in the form of flight, whereas sliding is a stepwise movement. In addition, the space exploration is much faster through flight. However, the precision is governed by the sliding motion. On that ground, it can be concluded that the condition of diversification and identification are accomplished by flying and sliding motion, respectively. The distance covered in a specific flight is a complex function of various parameters like the type and size of the mosquito, the nature of net, the nature of the pray, the time of the attack, presence of external force, and physical properties such as temperature and presence of light. Nonetheless, the distance covered in the subsequent flight may or may not be the same as shown in Fig. 1c. Similarly, the number of steps in a specific sliding cycle is also not constant. This is clearly shown in Fig. 1d. It shows that the distance covered in the sliding cycle which starts at point B may be ended by point C, C', or C". In addition to these, a mosquito can move its proboscis in the vicinity even if it is stationary, i.e., neither flying nor sliding.

In fact, the tendency of sliding and flying motion of mosquitoes searching for a hole is analogous to finding the optima of a complex multimodal function. This is different from simple MFO in the fact that the length of successive flight and sliding parameters are variable which results in fast convergence. The expansion and contraction parameters Phi1 and Phi2 are mathematically the Golden ratios, incorporating the features of Golden section search methods [24]. In addition, it includes the movement of proboscis which enhances the precision of the results. The algorithm is outlined as follows:

STEP 1 Initiate n particles.
STEP 2 Evaluate the fitness of the function.
STEP 3 Apply flying motion: Par = Phi1/Phi2 * par.
STEP 4 Evaluate the new particle.
STEP 5 Modify the fitness function if new fitness is better.
STEP 6 Apply sliding motion: Par = par ± J * C2.
STEP 7 Evaluate the new fitness.

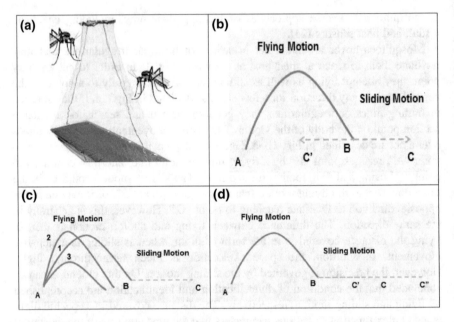

Fig. 1 Movements of mosquitoes. **a** Flying and sliding motion of mosquito. **b** Flying and sliding motion of a mosquito over a net. **c** Variable flying motion. **d** Variable sliding motion of a mosquito after a flight

STEP 8 Modify the fitness function if new fitness is better.
STEP 9 Apply precision movement Par = par $\pm$ C3 * par.
STEP 10 Evaluate the new fitness.
STEP 11 Modify the fitness function if new fitness is better.
STEP 12 Decrement iteration counter and evaluate the deviation. If iteration counter is not zero OR the deviation is less than DEVMIN, then go to STEP 3.
STEP 13 STOP.

3 The Experiment

The test of reliability, efficiency, and validation of the optimization algorithms is carried out on standard benchmarks or test functions [22, 23]. The present algorithm was tested on five benchmark functions of diverse nature, namely, Ackley, Rastrigin, Rosenbrock, Griewank, and Schwefel functions of 5, 10, and 30 dimensions.

3.1 Griewank Function

This is a continuous, differentiable, non-separable, scalable, and multimodal function. Its global minima lie at the origin. Mathematically, it is expressed by Eq. 1 [25].

$$f(x) = 1 + \frac{1}{4000} \sum_{i=1}^{d} x_i^2 - \prod_{i=1}^{d} \text{Cos}\left(\frac{x_i}{\sqrt{i}}\right) \tag{1}$$

3.2 Rosenbrock Function

The Rosenbrock function is continuous and is expressed by Eq. 2 [26]. It is a differentiable, non-separable, scalable, and unimodal function. Its global optimum is in a long, narrow, valley that curves parabolically along its axis resembling a banana and hence, also called the banana function. Its global minima is zero, which is located at unity.

$$f(x) = \sum_{i=1}^{d-1} \left[100\left(x_{i+1} - x_i^2\right)^2 + (x_i - 1)^2 \right] \tag{2}$$

3.3 Rastrigin Function

It is a non-convex, nonlinear, and multimodal function. Finding the minimum of this function is a difficult problem due to its large search space and large number of peaks. Mathematically, it is represented by Eq. 3 [27].

$$f(x) = 10d + \sum_{i=1}^{d} \left[x_i^2 - 10\cos(2\pi x_i) \right] \tag{3}$$

3.4 Ackley Function

This is a continuous, differentiable, non-separable, scalable, and multimodal function. This function has one narrow global optimum basin and numerous local optima and is considered to be a relatively trivial function due to its funnel shape. Nevertheless, this function has relevance for real-world applications since the free energy hypersurface of proteins is considered to be of similar, yet less symmetric, shape. It is represented by Eq. 4 [28].

$$f(x) = -20\exp\left(-2\sqrt{\frac{1}{d}\sum_{i=1}^{d}[x_i^2]}\right) - \exp\left(\sqrt{\frac{1}{d}\sum_{i=1}^{d}\cos(2\pi x_i)}\right) + 20 + \exp(1)$$

$$(4)$$

3.5 Schwefel Function

Schwefel's function has the difficulty of being less symmetric and having a global minimum at the edge of the search space. Additionally, there is no overall, guiding slope toward the global minimum like in Ackley's, or less extreme, in Rastrigin's function. It is expressed in Eq. 5 [9].

$$f(x) = 418.9829d - \sum_{i=1}^{d}\left[x_i\sin\left(\sqrt{|x_i|}\right)\right] \tag{5}$$

4 Results and Discussion

The computations were carried on a PC (Intel (R) Core(TM) i3, CPU @ 3.4 GHz and 4 GB RAM) in the Computer Application Laboratory of the Department of Petroleum Studies, using MATLAB 7.14. Thirty randomly generated seeds were provided for each function and dimension. The maximum iteration number (Maxit) was selected as the main loop termination criterion and its value was 200 for lower dimensions while 500 for higher dimensions. The flying, sliding, and precision constants, specifically Phi1, C1, and C2 were fixed at 1.618, 1.0, and 0.001, respectively. Finally, the population size was set at ten times the particle numbers. The outcome in terms of best value, mean value, and standard deviation (SD) for diverse functions are reported in Table 1. It can be observed that the algorithm is accurate and precise for lower dimensional problems. However, its performance deteriorates with growing dimensions. For instance, the best value and the mean value of 0000E+00 and 7.9800E−06, respectively, are obtained from the Griewank function, which are close to the ideal value. Further, the SD of 3.5393E−06 is ample testimony for the precision for the outcomes. Furthermore, it can be seen from the convergence analysis of Griewank function in Fig. 2a, that the final value arrived very fast, especially for lower dimensions. It converses to final value in two iterations for the five-dimensional problem (D5). A similar trend is observed for the Rastrigin function and, the Ackley function. However, for the Rosenbrock and

Table 1 Results on test functions

Function	Dimensions	Best value	Mean value	SD
Griewank function	5	0.0000E+00	7.9800E−06	3.5393E−06
	10	0.0000E+00	0.0000E+00	0.0000E+00
	30	1.9586E−03	5.4408E−03	2.1408E−03
Rastrigin function	5	0.0000E+00	2.7709E+00	2.6049E+00
	10	0.0000E+00	5.6785E+00	5.9619E+00
	30	0.0000E+00	3.4405E+01	3.8879E+00
Rosenbrock function	5	2.1800E−02	4.1800E−01	5.7000E−02
	10	1.0900E−01	6.3500E+00	1.2900E−02
	30	2.4783E+01	2.6479E+01	7.0024E−01
Ackley function	5	8.8818E−16	1.1548E−02	3.6026E−03
	10	8.8818E−16	8.8818E−16	9.8608E−32
	30	8.8818E−16	8.8818E−16	9.8608E−32
Schwefel function	5	2.0752E+03	2.0752E+03	9.3536E−02
	10	4.1504E+03	4.1504E+03	7.3887E−05
	30	1.2451E+04	1.2451E+04	1.7895E−03

Schwefel functions, the values are not reaching zero with increasing iterations, rather it maintains an almost constant offset.

The convergence analysis of all functions is depicted from Fig. 2a–e. From these analyses, it is observed that the difficulty of the conversion increases with dimensions. Figure 2c depicts the final value of the Rosenbrock function converges to its ideal value, which is zero. However, the difficulty of conversion increases as the dimension grows from five to thirty. Notwithstanding, the convergence of the Schwefel function is difficult. In fact, an almost flat profile is obtained, with an increasing gap for the higher dimensions as shown in Fig. 2e.

Tables 2, 3, 4, 5, and 6 provide a comparative analysis for the performance of the V-MFO algorithm with known algorithms for various functions. In Table 2, the results of the various algorithms on the Griewank function are given. For a five-dimensional problem, the best value of 0.0000E+00 and mean value of 7.9800E−06 is the lowest for the V-MFO among all the algorithms. Its results on the 10-dimensional problem are exceptionally good. However, for the 30-dimensional case, its performance (best value of 1.9586E−03) is lacking behind those of GA (1.9119 E−10), PSO (1.1102 E−15) and ABC (0.0000 E+00). Its performance in terms of mean value (5.4408E−03) is second best next to the ABC (6.3515 E−04). It can thus be concluded that the V-MFO performs very well on the Griewank function at lower dimensions. In Table 3, the comparative analysis of the various algorithms on the Rastrigin function is depicted. Here again, it can be observed that the best value is obtained in the V-MFO algorithm except the ABC algorithm.

From Table 4, it can be observed that the performance in terms of the best value is at the bottom for the five-dimensional Rosenbrock function. However, it is better

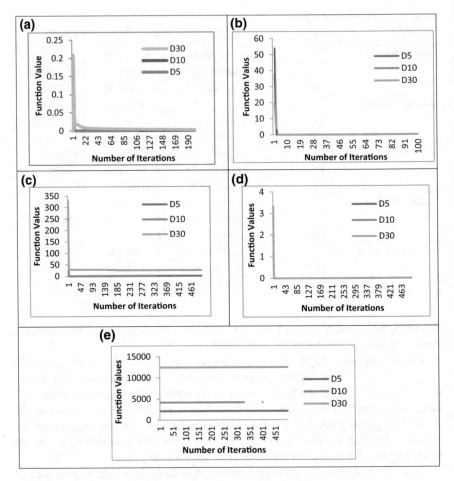

Fig. 2 a Convergence of Griewank function. **b** Convergence of Rastrigin function. **c** Convergence of Rosenbrock function. **d** Convergence of Ackley function. **e** Convergence of Schwefel function

than SLO in terms of the mean value. For the 30-dimensional Rosenbrock, the best value (2.4783E+01) and the mean value (2.6479E+01) are better than SLO but worse than the other test algorithms.

The comparative results for the various algorithms using the Ackley function are shown in Table 5. The V-MFO algorithm is better for 10- and 30-dimensional problem in all respects (i.e., in terms of best value, mean value, and SD). For the five-dimensional, the performance of the proposed algorithm is most desirable in terms of best value, but worst in terms of mean value. The values for the 5-, 10-, and 30-dimensional problems are higher than those obtained by other algorithms.

Table 6 depicts the comparative results of the various algorithms on the Schwefel function. The performance in terms of the best and the mean values are

Table 2 Comparative results on Griewank function

Algorithm	Dimension	Best value	Mean value	SD
V-MFO (proposed algorithm)	5	0.0000E+00	7.9800E−06	3.5393E−06
	10	0.0000E+00	0.0000E+00	0.0000E+00
	30	1.9586E−03	5.4408E−03	2.1408E−03
MFO	5	2.3368E−05	4.3400E−04	3.4695E−18
	10	3.3600E−04	5.8300E−04	3.2000E−05
	30	4.7400E−03	3.5000E−03	8.2900E−04
SLO	5	7.4000E−03	1.2870E−01	9.1200E−02
	10	2.4600E−02	3.1000E−01	5.2890E−01
	30	6.2000E−03	1.4705E+00	1.7894E+00
GA	5	1.2240E+01	1.2240E+01	1.2267E−10
	10	6.5624E+00	1.4568E+01	4.7149E+00
	30	1.9119E−10	1.4100E−02	2.4700E−02
PSO	5	7.4000E−03	2.3700E−02	1.2400E−02
	10	2.7000E−02	7.6900E−02	3.3600E−02
	30	1.1102E−15	1.3200E−02	1.4900E−02
ABC	5	0.0000E+00	6.8808E−04	2.1000E−03
	10	1.1102E−16	2.8000E−03	4.6000E−03
	30	0.0000E+00	6.3515E−04	3.5000E−03

Table 3 Comparative results of Rastrigin function

Algorithm	Dimension	Best value	Mean value	SD
V-MFO (proposed algorithm)	5	0.0000E+00	2.7709E+00	2.6049E+00
	10	0.0000E+00	5.6785E+00	5.9619E+00
	30	0.0000E+00	3.4405E+01	3.8879E+00
MFO	5	2.5500E−02	1.0400E+00	2.0000E−15
	10	2.2100E−01	1.7800E−01	2.1900E−01
	30	1.3300E+01	4.5500E+01	1.0500E+01
SLO	5	0.0000E+00	0.0000E+00	0.0000E+00
	10	9.9760E−01	2.2781E+01	1.4721E+01
	30	2.3147E+02	3.6123E+02	5.9041E+01
GA	5	2.0653E−09	6.3010E−01	6.1180E−01
	10	3.3258E−08	7.9600E−01	8.4260E−01
	30	2.5108E−06	3.0844E+00	2.3711E+00
PSO	5	0.0000E+00	6.6300E−02	2.5240E−01
	10	0.0000E+00	1.7267E+00	1.1662E+00
	30	1.7927E+01	2.9001E+01	8.5603E+00
ABC	5	0.0000E+00	0.0000E+00	0.0000E+00
	10	0.0000E+00	0.0000E+00	0.0000E+00
	30	0.0000E+00	1.0658E−15	4.2397E−15

Table 4 Comparative results of Rosenbrock function

Algorithm	Dimension	Best value	Mean value	SD
V-MFO (proposed algorithm)	5	2.1800E−02	4.1800E−01	5.7000E−02
	10	1.0900E−01	6.3500E+00	1.2900E−02
	30	2.4783E+01	2.6479E+01	7.0024E−01
MFO	5	2.1800E−02	4.1800E−01	5.7000E−02
	10	1.0900E−01	6.3500E+00	1.2900E−02
	30	3.2100E+01	4.4000E+01	1.0100E+01
SLO	5	1.6163E−08	1.3325E+00	8.7800E−01
	10	6.6976E−00	1.6880E+01	2.3646E+01
	30	2.5639E−03	1.6123E+04	9.1648E+03
GA	5	9.2000E−03	5.0800E−02	2.0100E−02
	10	1.0590E−01	7.0620E−01	5.6060E−01
	30	7.5000E−02	2.3368E−01	2.2251E+01
PSO	5	3.6080E−04	2.2040E−01	3.9500E−01
	10	3.9800E−02	2.6053E−00	1.3755E+00
	30	1.1177E+01	4.1055E+01	2.7793E+01
ABC	5	4.9000E−03	5.1300E−02	5.6700E−02
	10	1.9000E−03	5.2200E−02	5.1100E−02
	30	3.6401E−04	6.0700E−02	7.5300E−02

Table 5 Comparative results of Ackley function

Algorithm	Dimension	Best value	Mean value	SD
GR-VMFO (proposed algorithm)	5	8.8818E−16	1.1548E−02	3.6026E−03
	10	8.8818E−16	8.8818E−16	9.8608E−32
	30	8.8818E−16	8.8818E−16	9.8608E−32
MFO	5	1.3223E−02	9.4350E−02	1.3878E−17
	10	3.2106E−02	1.3715E−01	8.9775E−03
	30	2.4603E−01	5.1586E−01	1.2961E−01
SLO	5	−8.8818E−16	−8.8818E−16	0.0000E+00
	10	2.6645E−15	9.3400E−02	3.6120E−01
	30	6.2172E−15	9.2670E−01	2.3272E+00
GA	5	8.7429E−07	1.9147E−05	1.4740E−05
	10	4.1240E−06	2.8850E−05	1.2182E−05
	30	5.3960E−05	7.6502E−05	1.2228E−05
PSO	5	−8.8818E−16	2.0724E−15	1.3467E−15
	10	2.6645E−15	3.6119E−15	1.5979E−15
	30	1.5857E−09	1.7064E−08	2.1073E−08
ABC	5	2.6645E−15	2.6645E−15	0.0000E+00
	10	2.6645E−15	6.9278E−15	2.1681E−15
	30	1.6964E−13	4.4154E−13	3.0035E−13

Table 6 Comparative results of the Schwefel function

Algorithm	Dimension	Best value	Mean value	SD
GR-VMFO (proposed algorithm)	5	2.0752E+03	2.0752E+03	9.3536E−02
	10	4.1504E+03	4.1504E+03	7.3887E−05
	30	1.2451E+04	1.2451E+04	1.7895E−03
MFO	5	4.5400E−04	1.8700E−03	3.8500E−04
	10	6.0500E−03	1.5200E−02	4.4700E−04
	30	7.7700E−02	1.3300E−01	0.0000E+00
SLO	5	6.3638E−05	3.2505E+02	1.7285E+02
	10	7.5042E+02	1.2493E+03	2.7913E+02
	30	4.7145E+03	5.8700E+03	7.6059E+02
GA	5	2.0752E+03	2.0752E+03	5.2257E−11
	10	4.1504E+03	4.1504E+03	9.4875E−11
	30	1.2451E+04	1.2451E+04	7.3221E−10
PSO	5	6.3638E−05	2.9610E+02	1.3104E+02
	10	2.3688E+02	5.6719E+02	1.6709E+02
	30	1.5989E+03	2.8564E+03	4.0442E+02
ABC	5	6.3638E−05	6.3638E−05	1.9653E−14
	10	1.2728E−04	1.2728E−04	5.2425E−14
	30	3.8183E−04	3.1000E−03	1.4200E−02

inferior to those obtained by other algorithms. However, it is better than SLO and PSO in terms of SD.

Thus from the above discussion, it can be summarized that the V-MFO algorithm is comparatively more effective than other algorithms for the Griewank, Rastrigin, and Ackley functions, whereas its performance is comparatively less encouraging on the Rosenbrock and terrible on Schwefel function.

5 Conclusion

This paper presents a new algorithm variable mosquito flying optimization (V-MFO), which mimics the behavior of mosquitoes to find a hole in a mosquito net. The algorithm was tested on benchmark functions of various dimensions and modality, such as Ackley, Rastrigin, Rosenbrock, Griewank, and Schwefel functions of 5, 10, and 30 dimensions. It is concluded that the algorithm is more effective than other algorithms for Griewank, Rastrigin, and Ackley functions, whereas its performance is comparatively poor on Rosenbrock and Schwefel functions. The incorporation the golden ratios as the variable flying constant leads the result to more precise and accurate.

In a nutshell, the proposed algorithm is very effective in reaching the global optima for various diversified problems in general and extremely accurate for lower dimensional problems.

Acknowledgements The author is highly thankful to Dr. Rashid Ali, Associate Professor, department of Computer Engineering, for his invaluable guidance. In addition, many thanks to the faculty and staff of department of Petroleum Studies, Aligarh Muslim University, Aligarh, India for kind support for utilization of computational resources.

References

1. Mavrovouniotis M, Li C, Yang S (2017) A survey of swarm intelligence for dynamic optimization: algorithms and applications. Swarm Evol Comput 33:1–17
2. Kusakci A, Can M (2012) Constrained optimization with evolutionary algorithms: a comprehensive review. SouthEast Eur J Soft Comput 16–24
3. Goldberg DE, Holland JH (1988) Genetic algorithms and machine learning. Mach Learn 3 (2):95–99
4. Holland JH (1973) Genetic algorithms and the optimal allocation of trials. SIAM J Comput 2:88–105
5. Dorigo M, Caro GD (1999) The ant colony optimization meta-heuristic 11–32
6. Kennedy J, Eberhart R (1995) Particle swarm optimization. In: Proceedings of IEEE International Conference on Neural Networks, vol 4, pp 1942–1948
7. Pham DT, Ghanbarzadeh A, Koc E, Otri S, Rahim S, Zaidi M (2006) The bees algorithm–a novel tool for complex optimisation problems. In: Intelligent Production Machines Systems 2nd I* PROMS Virtual Conference 3–14 July 2006, p 454
8. Yang XS, Deb S (2009) Cuckoo search via Levy flights. In: Proceedings of 2009 World Congress on Nature Biologically Inspired Computing NABIC, pp 210–214
9. Wang P, Zhu Z, Huang S (2013) A new Meta-Heuristic technique for engineering design optimization: seven-spot ladybird algorithm, in 2nd International Symposium on Computer, Communication, Control and Automation, pp. 387–392
10. Passino KM (2010) Bacterial foraging optimization. Int J Swarm Intell Res 1:1–16
11. Yang XS (2012) Flower pollination algorithm for global optimization. In: Lecture notes in computer science (including subseries lecture notes in artificial intelligence and lecture notes in bioinformatics), pp 240–249
12. Yazdani M, Jolai F (2015) Lion optimization algorithm (LOA): a nature-inspired metaheuristic algorithm. J Comput Des Eng 3:1–14
13. Mucherino A, Seref O (2007) Monkey search: a novel metaheuristic search for global optimization. In: AIP conference proceedings, pp 162–173
14. Yang XS, Deb S (2010) Eagle strategy using Levy walk and firefly algorithms for stochastic optimization. In: Studies in computational intelligence, pp 101–111
15. Mirjalili S, Mirjalili SM, Lewis A (2014) Grey wolf optimizer. Adv Eng Softw 69:46–61
16. Mirjalili S, Lewis A (2016) The whale optimization algorithm. Adv Eng Softw 95:51–67
17. Gandomi AH, Alavi AH (2012) Krill herd: a new bio-inspired optimization algorithm. Commun Nonlinear Sci Numer Simul 17:4831–4845
18. Javidy B, Hatamlou A, Mirjalili S (2015) Ions motion algorithm for solving optimization problems. Appl Soft Comput J 32:72–79
19. Alauddin M (2016) Mosquito flying optimization (MFO). In: 2016 international conference on electrical, electronics, and optimization techniques (ICEEOT). IEEE, pp 79–84
20. Youssef H, Sait SM, Adiche H (2001) Evolutionary algorithms, simulated annealing and tabu search: a comparative study. Eng Appl Artif Intell 14:167–181

21. Lehane MJ (2009) Blood sucking. In: Encyclopedia of insects, pp 112–114
22. Harbach RE, Besansky NJ (2014) Mosquitoes. Curr Biol 24(1)
23. Quiroz-Martínez H, Rodríguez-Castro A (2007) Aquatic insects as predators of mosquito larvae. J Am Mosq Control Assoc 23:110–117
24. Avriel M, Wilde DJ (1968) Golden block search for the maximum of unimodal functions. Manag Sci 14:307–319
25. Locatelli M (2003) A note on the Griewank test function. J Glob Optim 25:169–174
26. Shang Y-W, Qiu Y-H (2006) A note on the extended Rosenbrock function. Evol Comput 14:119–126
27. Mühlenbein H, Schomisch M, Born J (1991) The parallel genetic algorithm as function optimizer. Parallel Comput 17:619–632
28. Williams RJ, Peng J (1991) Function optimization using connectionist reinforcement learning algorithms. Conn Sci 3:241–268

Conclusion

Rashid Ali and M. M. Sufyan Beg

In this chapter, we summarize the main contributions of different chapters in the book. The book consists of 14 contributed chapters, which are organized in many different parts.

The first part comprising of four chapters, second to fifth chapters, is devoted to soft computing based recommender systems. In chapter entitled "Context Similarity Measurement Based on Genetic Algorithm for Improved Recommendations", authors proposed to use the genetic algorithm to improve the performance of context-aware fuzzy collaborative filtering based recommender systems. They evaluated the proposed approach using two parameters, namely mean absolute error (MAE) and coverage. They experimented with real-world movie LDOS-CoMoDa dataset and compared the results of the proposed approach with three state-of-the-art filtering approaches, namely, Pearson collaborative filtering (PCF), fuzzy collaborative filtering (FCF), and context-aware fuzzy collaborative filtering (CA-FCF). Results of the comparison clearly confirmed that the proposed approach outperformed the other three existing techniques with least value of MAE and highest value of coverage. Hence, they concluded that the use of the genetic algorithm is a promising way to improve the effectiveness of collaborative filtering based recommender system.

In chapter entitled "Enhanced Multi-criteria Recommender System Based on AHP", the authors proposed to use analytic hierarchy process (AHP) for multi-criteria decision-making based college recommender system. They experimented with a sample dataset collected from U.S. News dataset. The dataset consisted of ratings on various aspects of many different colleges from multiple users.

R. Ali (✉) · M. M. S. Beg
Department of Computer Engineering, Z. H. College of Engineering and Technology,
Aligarh Muslim University, Aligarh 202002, Uttar Pradesh, India
e-mail: rashidaliamu@rediffmail.com

M. M. S. Beg
e-mail: mmsbeg@eecs.berkeley.edu

© Springer Nature Singapore Pte Ltd. 2017
R. Ali and M. M. S. Beg (eds.), *Applications of Soft Computing for the Web*,
https://doi.org/10.1007/978-981-10-7098-3_16

They used eight criteria and eight colleges. The authors validated their approach using two parameters, average precision and Kendall tau rank correlation coefficient. The experimental results with high value of average precision and Kendall tau rank correlation coefficient clearly demonstrated the effectiveness of the use of AHP in the college recommendations.

In chapter entitled "Book Recommender System Using Fuzzy Linguistic Quantifiers", the authors presented a book recommender system based on the aggregated ranking of books, which was obtained by the aggregation of different rankings of the books given by respective universities using ordered weighted aggregation (OWA), a well-known fuzzy aggregation operator. To evaluate the performance of their system, they obtained the true ranking of the books on the basis of experts' feedback. They used three parameters P@10, FPR@10 and mean average precision (MAP) to evaluate their method and compared the performance of their system with Amazon. Experimentally, they showed that the results of the recommender system using linguistic quantifier "at least half" were better than that of Amazon and other two systems used in the experiments.

In chapter entitled "Use of Soft Computing Techniques for Recommender Systems: An Overview", the authors discussed different studies performed on soft computing based recommender systems. They summarized perceptively and comprehensively the recent research on recommender systems using the soft computing techniques such as fuzzy sets, neural networks, evolutionary computing, and swarm intelligence. This survey covered recent recommendations approaches separately for each soft computing technique and outlined several possible research directions for each technique.

The second part consisting of sixth chapter is devoted to soft computing based online documents summarization. Chapter entitled "Hierarchical Summarization of News Tweets with Twitter-LDA", discussed the hierarchical summarization of news tweets. Here, the main topics in the tweet dataset were found using modified Twitter-LDA topic model with tweet pooling based on hashtags and replies. Tweet summary for each topic was generated by finding important tweets using retweet and reply count and change in word distributions along the temporal dimension. Precision and recall performance measures were used to analyze the performance of the application. Experimental results clearly demonstrated that extractive summary can be produced with satisfactory precision and recall values.

The third part consisting of seventh chapter is devoted to soft computing based web data extraction. In chapter entitled "Bibliographic Data Extraction from the Web Using Fuzzy-Based Techniques", the authors discussed the bibliographic data extraction from the Web. They have used DBLP (Digital Bibliography & Library Project) live dataset of publications from the field of computer science and obtained results using proposed fuzzy data extraction tool. Results of their experimentation depicted that the proposed technique extracted live publication data, which has clearly the advantage of performing better over static datasets.

The fourth part consisting of eighth chapter is devoted to soft computing based question answering systems. In chapter entitled "Crop Selection Using Fuzzy Logic-Based Expert System", the authors presented fuzzy logic based system, which can be used for crop selection depending upon the given climatic conditions and soil properties. The proposed system is customizable. Hence, any number of new crop rules can be added to the system. The input parameters were taken for Indian region. The authors also proposed to provide user-friendly GUI system which will be of great help to the farmers in general. They also pointed out that the system can be extended to include different clauses for crops in growing phase and in harvest period. These clauses can contain pest diagnosis for each crop and harvesting tips too.

The fifth part consisting of ninth chapter is devoted to soft computing based online healthcare systems. Chapter entitled "Fuzzy Logic Based Web Application for Gynaecology Disease Diagnosis", discussed the development of fuzzy logic based web application for disease diagnosis. The proposed Medical Decision Support System (MDSS) addresses perception-based diagnosis of patients for gynecological diseases. This method might be suitable for other MDSS as well, with suitable modifications.

The sixth part consisting of tenth chapter is devoted to soft computing based online documents clustering. In chapter entitled "An Improved Clustering Method for Text Documents Using Neutrosophic Logic", the authors proposed improved fuzzy c-means clustering algorithm based on neutrosophic logic. The authors compared the performance of the proposed algorithm with the traditional fuzzy c-means algorithms using three evaluation parameters precision, recall and F-measure. Experimentally, the authors deduced that the proposed improved fuzzy c-means clustering algorithm using neutrosophic logic outperformed the traditional fuzzy c-means clustering algorithm.

The seventh part consisting of eleventh chapter is devoted to soft computing based web security applications. In chapter entitled "Fuzzy Game Theory for Web Security", authors proposed an algorithm for web security with fuzzy game theory approach. The authors performed a simulation with 1000 sample of attacks to measure the performance of the proposed method and presented the results of this simulation. The results of simulation clearly established the author's claim that the proposed approach is able to tackle any random unauthorized data access to a large extent.

The eighth part consisting of twelfth chapter is devoted to soft computing based online market intelligence. In chapter entitled "Fuzzy Models and Business Intelligence in Web-Based Applications", the authors discussed business decision-making with vague data in presence of outliers. The authors concluded that the outliers, which have strong effect on an important step, are known as hotspot outliers and they can cause incorrect decisions. They attempted to distinguish algorithms for their capability to normalize the impact of hotspot outliers with an emphasis that irrespective of this capability an algorithm may still be the desired one. Therefore, authors proposed an outline of the algorithm that can be used as preprocessing step to normalize the impact and for making more appropriate decision. The authors also presented detailed design aspects of web-based application where decision-making and fuzzy logic have an important place and finally proposed a novel design for the cloud-based solution.

The ninth part consisting of thirteenth and fourteenth chapters is devoted to soft computing based Internet of Things applications. In chapter entitled "GSA-CHSR: Gravitational Search Algorithm for Cluster Head Selection and Routing in Wireless Sensor Networks", the authors tried to address the two most promising problems in wireless sensor networks (WSNs) namely cluster head selection and routing using an optimization algorithm called gravitational search algorithm. The proposed algorithm (GSA-CHSR) was extensively tested with existing techniques on various scenarios of the network to study the performance. They experimented to compare the performance of the proposed algorithm with five existing methods such as DHCR, EADC, hybrid routing, GA-routing and PSO-routing using five test metrics namely residual energy, number of alive nodes, network lifetime, data packets received at the base station, and convergence rate. The experimental results clearly demonstrated the superiority/competitiveness of GSA-CHSR over the other five existing techniques.

In chapter entitled "Utilizing Genetic-Based Heuristic Approach to Optimize QOS in Networks", the author proposed a genetic algorithm based approach to optimize routing related issues and QOS parameters like hop count, delay, throughput, etc. The author performed simulation to measure the performance of the proposed approach and Dijkstra's algorithm based approach and presented the results of the simulation. The results of simulation deduced that heuristic-based genetic approach led to neoteric optimized results, found fittest shortest path and optimized QOS performance parameters. The proposed approach also outperformed the Dijkstra's algorithm based approach.

The tenth part consisting of fifteenth chapter is devoted to other emerging soft computing techniques and applications. In chapter entitled "V-MFO: Variable Flight Mosquito Flying Optimization", the author presented a new optimization scheme called Variable Flight Mosquito Flying Optimization. The proposed optimization algorithm was tested for the global minima on diverse types of benchmark functions of diverse dimensions and modality; such as Ackley, Griewank, Rastrigin, Rosenbrock, and Schwefel functions of 5, 10, and 30 dimensions. The results were compared with five established methods, namely genetic algorithm, particle swarm optimization, seven-spot ladybird optimization, artificial bees' colony, and mosquito flying optimization. Results of the comparison confirmed that the proposed algorithm outperformed the other existing algorithms.

We conclude with the statement that, in this book, we have just tried to present some of the soft computing applications for the Web. This field is so important and diverse that many other good books can be written and presented before the world in the future.

Printed in the United States
By Bookmasters